Why buy a Ho Math Chess Franchise?

Intellectual Properties of Ho Math Chess

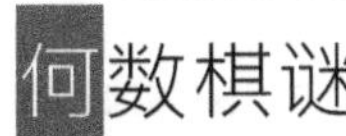

- www.homathchess.com

Student's name: ________________ Assignment date: ___________________

Table of Contents

Why buy a Ho Math Chess Franchise?

Intellectual Properties of Ho Math Chess

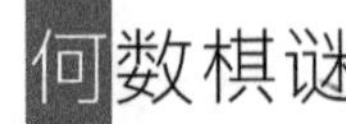

- www.homathchess.com

Student's name: ________________ Assignment date: ____________________

Student's name: ________________ Assignment date: ____________________

Student's name: ________________ Assignment date: ____________________

Preface

Many people have asked us about our franchise business and would like to know more about it before making their decisions on opening a Ho Math Chess franchised learning centre. They are interested in finding out more about our workbooks' uniqueness and their effectiveness in teaching. They do not know how our innovative and revolutionary math, chess and puzzles combined workbooks are different from traditional math workbooks. Many of them also wonder if our materials will meet their own countries' math curriculums. All of these are legitimate questions. There is no better way than that we let our potential franchisees have the privilege to eyewitness our workbooks and do due diligence by researching our teaching methodology and philosophy, our workbooks, and finding out why our centre is a fun math learning centre. This sincere divulgence of our sensitive business information helps potential franchisees make a wise decision as to why buying Ho Math Chess instead of others.

Ho Math Chess franchise business centres on its innovative teaching methodology using math, chess, and puzzles integrated workbooks created by using our copyright-protected property intellectual properties. Our business has a substance that differentiates fundamentally from other competitors. In this book, we will address these properties. Our flagship invention Frankho ChessDoku also has won book awards.

Some of the franchisees would like to find out how we integrated math, chess, and puzzles in one workbook and wonder if they make learning math more exciting and fun for kids. Answers for all these questions could be found in this book, and also a sample of our worksheets are provided. The referrals can witness our teaching's effectiveness from our parents, and those referrals can be read from our website www.mathandchess.com. The USA Illinois research paper has shown that the Ho Math Chess teaching method statistically significantly increases children's math marks and improves children's critical thinking skills.

We would like to show that franchisees are not just buying our business know-how; they are buying our innovative materials and using our inventions. We have many proprietaries and internationally copyright protected intellectual properties which other similar math learning centres could not offer.

We hope that reading through this book will allow you to understand how unique our quality workbooks are and how fun our teaching methodology is. Ho Math Chess is genuinely a fun math learning centre, and it truly helps kids raise their math marks and improve their thinking skills.

Many articles included here were previously published by Frank and Amanda Ho. Ho Math Chess will be expanding business in China, so some Chinese articles are also included in this book. Some Spanish articles are also included here to help potential franchisees in Spanish speaking countries.

Frank Ho

Amanda Ho
February 2014

Student's name: ________________ Assignment date: ____________________

Cover design description - Ho Math Chess Geometry Chess Symbol

Ho Math Chess founder Frank Ho invents the following copyrighted Geometry Chess Symbols. They are being used as an innovative way of linking chess and math to create math and puzzle workbooks other than being used as a teaching chess set. These symbols are static. They can only become lively commanders when Ho Math Chess used these symbols to have invented and created the Symbolic Chess Language (Details see the Symbolic Chess Language section.) By using the Symbolic Chess Language, students can perform arithmetic operations.

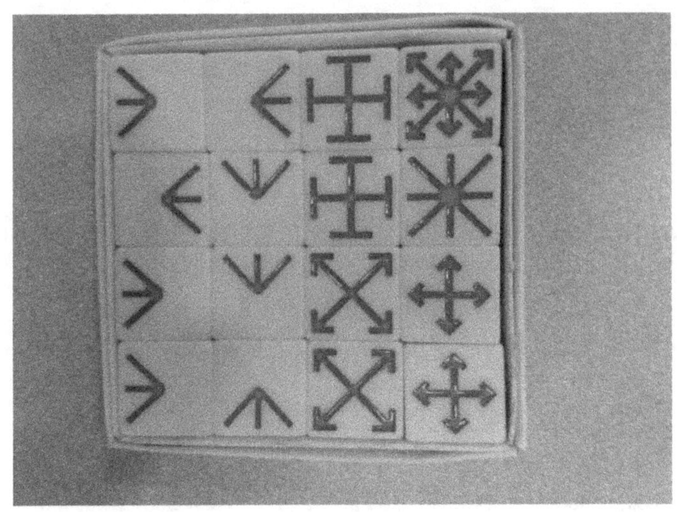

Student's name: ________________ Assignment date: ____________________

How did Ho Math Chess™ get started?

Frank Ho, a certified math teacher, was intrigued by math and chess relationships after teaching his son chess started **Ho Math Chess™** in 1995. His long-term devotion to research has led his son to become a FIDE chess master and Frank's publications of over 20 math workbooks. Today **Ho Math Chess™** is the world's largest and only franchised scholastic math, chess, and puzzles specialty learning centre worldwide. **Ho Math Chess™** is a leading research organization in math, chess, and puzzles integrated teaching methodology.

There are hundreds of articles already published showing chess benefits children and that math puzzles are an excellent way of improving brainpower. The learning effect is more significant by integrating chess and mathematical puzzles to spark interest in learning math.

Parents send their children to **Ho Math Chess™** because they like Ho Math Chess™ teaching philosophy – offering children problem-solving questions in various formats. The questions could be pure chess, chess puzzles, mathematical puzzles like logic, pattern, three structures, Venn diagram, probability and many more math concepts.

Ho Math Chess has developed a series of unique and high-quality workbooks. Its first product is the world's first **Magic Chess and Fun Math Puzzles**. The workbook is not only for learning chess but also for enriching math ability. Right from the creation time, **Ho Math Chess** has set itself apart from other math learning centres, chess clubs or chess classes.

The purposes of **Ho Math Chess™** teaching method and workbooks are to:

- improve math marks.
- develop problem-solving and critical thinking skills.
- improve strategic thinking ability.
- boost brainpower.

Testimonials, sample worksheets, and franchise information can found at www.mathandchess.com.

Below is a brief of Ho Math Chess Learning Centre's history.

1995 - **Frank Ho founded Ho Math Chess Learning Centre**. It was located at #5, 5729 West Boulevard, Vancouver, with one classroom only.

1996

- **Ho Math Chess** moved to a bigger location. It was located at #12, 5729 West Boulevard, with three classrooms.
- **Ho Math Chess** program started to offer at St. George's Summer School in Vancouver.

Student's name: ________________ Assignment date: ____________________

1997

- **Mathematical Chess Puzzles for Junior** officially published.

1998 - **Fun Math Puzzles for Juniors** officially published.

1999 - **Ho Math Chess** moved to a location with exposure to the 3-storefront with five classrooms. Rooms #4 and # 6 at 2265 West 1st, Vancouver, was covered with dust when Frank Ho discovered it. Frank applied for a city license and arranged an inspection, and negotiated the lease to secure the place.

2000, 2001

Ho Math Chess math team coached by Frank Ho competed in the Mathcounts and won the third in the most competitive Vancouver region. The Ho Math Chess team was not allowed to compete provincially and forever banned from Mathcounts competition using a learning centre's name because of performance. (Note: this policy had since been relaxed in 2003.)

2003 - **Math Contest Preparation** was officially published.

2004

- Whole Numbers Operations - addition, subtraction published.
- Whole Numbers Operations - multiplication, division published.
- Whole Numbers Problem Solving and Puzzles published.
- Fractions, Decimals, Ratio, Equations published.
- Pre-Algebra Problem Solving Strategies published.
- **Ho Math Chess** franchise business worldwide launched.
- Franchisees Richmond, Burnaby, Ecuador joined **Ho Math Chess** Learning Centre
- **Ho Math Chess** was trademarked.

2005

- **Magic Chess Math Puzzles** 3rd generation workbook published.
- Creation of the **Magic Chess and Math Puzzles Prime** for pre-schoolers.
- **Ho Math Chess** trademark was approved.

2006

- Article on **Enriching Math Using Chess** approved for publishing.
- Peninsula **Ho Math Chess** opened in California, USA
- Palo Alto **Ho Math Chess** opened in California, USA
- Chicago, Illinois **Ho Math Chess** opened, USA
- Frank gave a presentation on Chess in School Curriculum to Surrey teachers in Canada on Pro-D day.
- Many worldwide franchisees joined Ho Math Chess this year.

2007

- Master Franchise in Latin America awarded.
- Ho Math Chess continues to improve products, and this year is the breakthrough year in Ho Math Chess' growth history. Many new inventions and products are produced.
- Many worldwide franchisees have joined Ho Math Chess.

2008

- The master franchise of all Arabic speaking countries awarded.
- The master franchise of Singapore was awarded.
- The master franchise of Malaysia was awarded.
- Franchise in Taiwan awarded.
- Franchise in Spain awarded.
- Franchise in Peru awarded.
- Frankho Chess Mazes invented and is running as a regular column in Vancouver Chinese paper.
- IQ Chess Brainpower workbook created for children.

Student's name: ________________ Assignment date: ____________________

2009

- Franchise in Mexico City awarded.
- Brainpower workbook produced
- Copyright of Geometry Chess Symbol awarded in Canada.
- Trademark of Geometry Chess Symbol awarded in Canada.

2010

- The additional new centre of Singapore Ho Math Chess opened.
- Some Ho Math Chess workbooks start to sell worldwide in March.
- Frankho Puzzles, including the invented Geometry Chess Symbols trademark, has been approved in Canada.
- The additional new centre of Texas Ho Math Chess opened.
- Awarded franchise in Nigeria.

2011

- Released the world's first Chinese Ba Gua math for children.
- Published world's first math, chess, and Sudoku puzzles for children workbook on www.amazon.com.
- Releases of workbook Grade 8 and Grade 9.
- Malaysia joins Ho Math Chess.
- Starts to produce Ho Math, Chess, and Puzzles worksheets for the chess club.

2012

- Published math workbook for preschoolers/kindergarteners on www.amazon.com.
- Filed trademark in China.
- Successfully field tested worksheets for chess and puzzles club.
- Completion of Malaysia franchise training.
- Malaysia Ho Math Chess grand opening launched.

2013

- Ho Math Chess workbooks translated into Turkish.
- Ho Math Chess Chinese trademark 何数棋谜 has been approved in China.

2014

- Many new Ho Math Chess workbooks are published.
- Malaysia Ho Math Chess second location is open.

2020

- Ho Math Chess starts an online teaching business model.

Why buy a Ho Math Chess Franchise?

Intellectual Properties of Ho Math Chess

何数棋谜 - **www.homathchess.com**

Student's name: ________________ Assignment date: ____________________

For area codes: (613) (519) (705) call: 1-800-387-9022

Montreal: Steve LeClair (514) 982-3050
For area codes: (514) (819) (418) **call:** 1-800-368-7526

CANADA'S NATIONAL NEWSPAPER
THE GLOBE AND MAIL
A partner in building your business.

Saturday, November 25, 1995

Chess

Vancouver lad lands fifth place

BY JONATHAN BERRY
Special to The Globe and Mail
Nanaimo, B.C.

Andrew Ho of Vancouver placed fifth in the world under-12 championship in Sao Lourenco, Brazil. He tallied 7.5 points out of 11.

Etienne Bacrot took first place with 10 points, including a win over Ho, and two points ahead of the closest competition. Bacrot is so highly regarded in his native France that he was given a special place in the Paris PCA Grand Prix. Vladimir Kramnik, one of the top players in the world of any age, defeated him 1.5 to .5, and Bacrot did not look out of place until late in the second game.

The annual World Youth Chess Festival now comprises world championships for ages under-10, -12, and -14, for both boys and girls. Canada holds its age-group championships without regard to gender, and as the champions were all boys, they took part in the boys sections. Had any of our champions been a girl, she could have elected the section in which to play. The national competition, and the expenses of going to Brazil, were sponsored by the Ottawa computer software company, Corel.

Two players from Quebec, Marc Fortin of St-Laurent in the under-10, and Eugene Cormos of Montreal in the under-14, both scored a creditable 50 per cent.

Unlike Canada, the Netherlands sent only representatives to the under-12, one boy and one girl. Azadmanesh Moosa, of that country, had White against Ho in the last round.

1.e2-e4 c7-c5 2.Ng1-f3 e7-e6 3.d2-d4

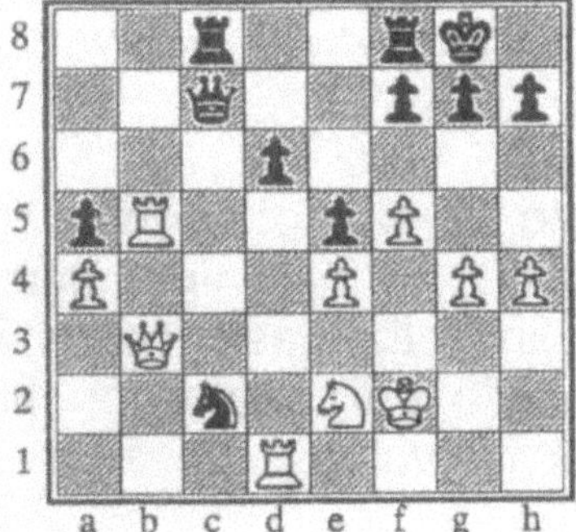

Black to play.

c5xd4 4.Nf3xd4 Ng8-f6 5.Nb1-c3 d7-d6 6.Bf1-c4 Bf8-e7 7.f2-f4 O-O 8.Bc4-b3 Nb8-a6

The old saw "Knight on the rim, prospects are dim" does not apply here because the knight quickly gets to c5.

9.Qd1-f3 Na6-c5 10.f4-f5 a7-a6 11.g2-g4

Too ambitious. If he wanted to push this pawn to g5, he should have held the move f4-f5 in reserve. Now White's dark squares become tender.

11. . . . Nf6-d7!

Heading for e5. Already, Black has for his knights two beautiful central posts, which outweigh White's space advantage.

12.Bc1-e3 Nd7-e5 13.Qf3-g2 Be7-h4+ 14.Ke1-d2 Bh4-g5!

Although adventures may have seemed appealing, Black tends to business first. The player who is cramped does well to trade pieces, especially those that defend the squares left vulnerable by pawn advances. The text also liquidates White's threat, which was g4-g5.

15.Ra1-e1 b7-b5 16.h2-h4 Bg5xe3+ 17.Re1xe3 b5-b4 18.Nc3-e2 a6-a5 19.a2-a4 b4xa3 *en passant* **20.b2xa3 Nc5xb3+ 21.Re3xb3 Ne5-c4+ 22.Kd2-e1 e6-e5 23.Nd4-b5 Bc8-a6 24.a3-a4 Ba6xb5!**

An astute move. If 25.a4xb5, Black wins with 25. . . . a5-a4 and 26.Rb3-b4 Nc4-e3, or 26.Rb3-c3 Qd8-a5.

25.Rb3xb5 Nc4-e3 26.Qg2-f3 Ne3xc2+ 27.Ke1-f2 Ra8-c8

Or 27. . . . Qd8-c7, eyeing Qc7-a7+ and intending to bring the other rook into play on the queenside.

28.Qf3-b3 Qd8-c7 29.Rh1-d1

DIAGRAM: **29. . . . Nc2-d4!**

Black returns the pawn to open up an ambush on White's king. Alternatives include Nc2-b4 or Rc8-b8, but the text gets to the heart of the position.

30.Ne2xd4 e5xd4 31.Qb3-d3?

Too subtle. After 31.Rd1xd4 d6-d5, White has better chances of survival than in the game.

31. . . . d6-d5 32.Rb5xd5 Qc7-f4+ 33.Kf2-g2 Qf4xg4+ 34.Kg2-f2 Qg4xh4+ 35.Kf2-e2 Qh4-h2+ 36.Ke2-e1 Rc8-c3 37.Qd3xd4 Rc3-g3 38.Ke1-f1 Qh2-h3+ 39.Kf1-e1 Rg3-g2 40.Qd4-f2 Rg2xf2 41.Ke1xf2 h7-h6 42.Rd1-d3 Qh3-h4+ 43.Kf2-f3 Rf8-e8 44.Rd3-e3 g7-g5 45.f5xg6 *en passant* **f7xg6 46.Rd5xa5 Re8-f8+ 47.Kf3-e2 Qh4-f2+ 48.Ke2-d3 Qf2-f1+ 49.Kd3-d4 Qf1-a1+ 50.Re3-c3 Rf8-c8**

White resigned. Ho is 12 years old — the "under-12" refers to the player's age on Jan. 1, 1995. Most master-class players of any age would be happy to play as focused a game as Ho did here.

Student's name: ________________ Assignment date: ____________________

What is the mission of Ho Math Chess?

Ho Math Chess' mission is to challenge each student's potential to the fullest and excel each student's problem-solving ability to the highest by providing a unique and enriched mathematics curriculum.

Ho Math Chess is an after-school math specialty learning centre, not a specialized chess centre. Ho Math Chess has a mission to promote math learning and help children improve or advance their math knowledge beyond their school curriculum levels. Ho Math Chess uses chess and puzzles and integrates them into math workbooks to help children learn math, improve their brain power, and spark their math interests. The consequence is children will be more interested in problem-solving, become smarter, know how to play chess, and solve puzzles.

Ho Math Chess does not vigorously promote the idea of asking children to play chess. So if a child has no interest in learning chess, then it is perfectly fine because to work on Ho Math Chess workbooks only requires children to know the basic moves of chess pieces. No chess tactics or strategies are required. It takes only a few minutes for children to grasp the knowledge of chess moves and their respective values.

Student's name: ________________ Assignment date: ___________________

Why is Ho Math Chess™ a stand-out learning centre?

Ho Math Chess creates the world's first commercially available math and chess integrated workbook using its invention, and internationally copyright protected Ho Math Chess Geometry Chess Symbols and its Ho Math Chess teaching set by using these symbols.

GCS (Geometry Chess Symbols) is used in Ho Math Chess workbooks to link chess, math, and puzzles. GCS also has created an award-winning workbook Frankho ChessDoku, Frankho ChessMaze, and many other Ho Math Chess' math, chess, and puzzles integrated workbooks.

Ho Math Chess is a math specialty after-school learning centre emphasizing its fun math teaching methodology using integrated problem solving and math IQ puzzles workbooks.

Ho Math Chess is not just selling the idea of teaching math and chess under the same roof. We are offering integrated math, chess, and puzzles printed workbooks. In brief, Ho Math Chess has the following unique intellectual properties:

- Trade name and trademark: Ho Math Chess
- The integrated teaching methodology
- Workbooks creator and copyrights holder of the following workbooks and inventions:
  - GCS
  - Integrated math, chess, and puzzles fun workbooks using GCS.
  - Ho Math Chess teaching chess set
  - Inventions of *Frankho ChessDoku*, and *Frankho ChessMaze* and many other unique workbooks such as *Math Contest Preparation*, and *Problem Solving and Math IQ Puzzles*

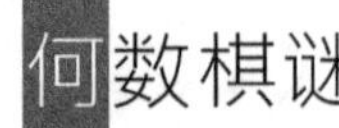

Student's name: ________________ Assignment date: ____________________

Why are Ho Math Chess workbooks unique?

Many math educators and research papers have shown that the idea of using a game would be a fun and effective way of teaching math. Some guidelines or lesson plans are also available for teachers to conduct a few lessons. No full-blown curriculum-based chess, puzzles, and math integrated course materials ever were developed in the past until Mr. Frank Ho took action to create the math, chess, and puzzles integrated and curriculum-based workbooks. This teaching model and methodology has been supported by a publication of over 50 research papers and over 30 workbooks in the past 20 years of Ho Math Chess' teaching and research. This is why Ho Math Chess workbooks are unique and are disruptively different from the traditional workbooks.

Ho Math Chess is the world's first and the only commercially available establishment globally, with over 20 years in math, chess, and puzzles integrated and printed workbooks to teach children from pre-school to elementary school. No longer teachers have to rely on a few scattered chess lesson plans to conduct math courses. Instead, math teachers can use Ho Math Chess integrated workbooks to teach children math. Ho Math Chess workbooks are one-of-its-kind, unique, and copyrighted math, chess, puzzles truly integrated workbooks (Canada copyright no. 1069744, Chess symbols Trademark TMA771400).

Ho Math Chess materials also have different levels to suit students with different backgrounds. So for gifted students, they can use Ho Math Chess math contest workbooks. For remedial students, they can use fast-track workbooks.

Ho Math Chess could not create this math, chess, and puzzles integrated workbooks without using its invention that is internationally copyrighted Geometry Chess Symbol. By using this invention, math and chess are linked, and students can do math using mini puzzle-like problems. Working on Ho Math Chess worksheets, students become more interested in math than working on traditional math worksheets.

A sample of Ho Math Chess workbooks

Ho Math Chess has published over 30 unique math, chess, and puzzles workbooks. Below is a collection of cover images of some workbooks.

Test of Future Math Star	**Pre-K and Kindergarten Math**
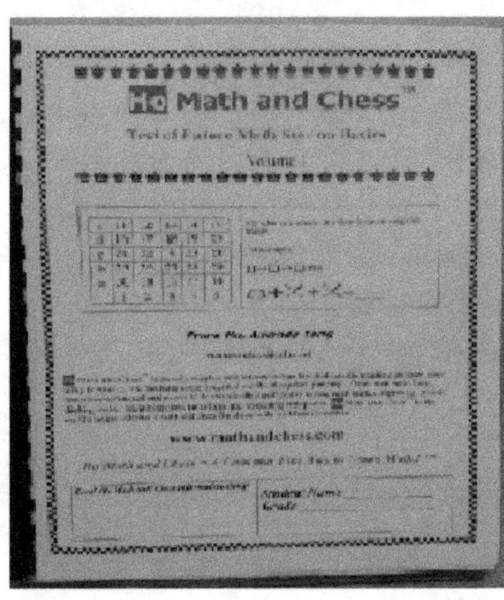	
Kindergarten Math	**Learning Calculation without Using Fingers**

Student's name: ________________ Assignment date: ____________________

Addition	**Subtraction**
	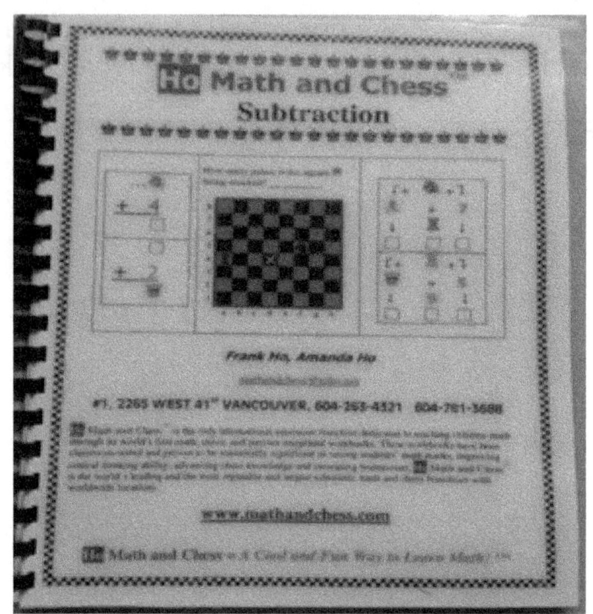
Multiplication	**Division**

Student's name: ________________ Assignment date: ___________________

Addition and Subtraction	Whole Number Operations

Fundamental Math	Problem Solving and Math IQ Puzzles

Student's name: ________________ Assignment date: ________________

Principle Math	Frankho 3 by 3 ChessDoku

Frankho 4 by 4 ChessDoku	Ho Math Chess Sudoku IQ Puzzles Sample Worksheets

Student's name: ________________ Assignment date: ____________________

Basics Whole Numbers

Test Future Math Stars on Basics

Fast Track Computations

Math 9

Student's name: ________________ Assignment date: ____________________

Pre-calculus 10	High-Performance Math on Basic Operations
	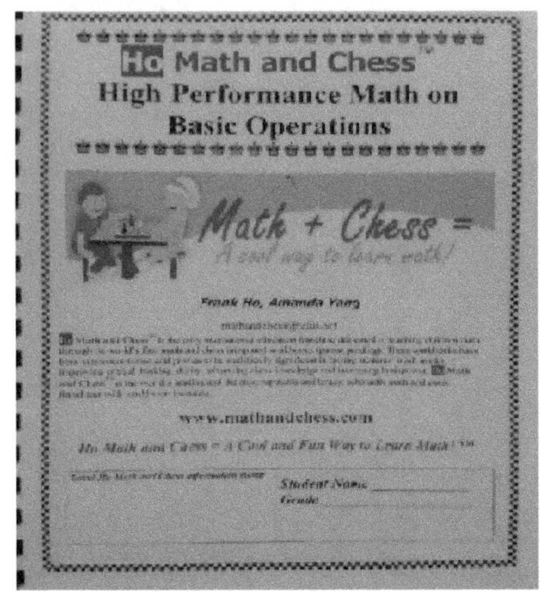

Student's name: ________________ Assignment date: ___________________

Ho Math Chess™

Math Contest Preparation

8	3	4	8	1	6	6	7	2	6	1	8
1	5	9	3	5	7	1	5	9	7	5	3
6	7	2	4	9	2	8	3	4	2	9	4
2	9	4	2	7	6	4	9	2	4	3	8
7	5	3	9	5	1	3	5	7	9	5	1
6	1	8	4	3	8	8	1	6	2	7	6

Frank Ho, Amanda Ho

mathandchess@telus.net

Ho Math Chess™ is the only international education franchise dedicated to teaching children math through its first math, chess, and puzzles integrated workbooks. These workbooks have been classroom-tested and proven to be statistically significant in raising students' math marks, improving critical thinking ability, advancing chess knowledge and increasing brainpower. **Ho Math Chess**™ is the world's leading and most reputable and largest academic math and chess franchiser worldwide.

www.homathchess.com

Ho Math Chess = ***A Cool and Fun Way to Learn Math!*** ™

Student's name: ________________ Assignment date: ____________________

♚♚♚♚♚♚♚♚♚♚♚♚♚♚♚♚♚♚♚♚♚♚

Ho Math Chess™

Fundamental Magic Chess and Math Puzzles

♚♚♚♚♚♚♚♚♚♚♚♚♚♚♚♚♚♚♚♚♚♚

For Problem Solving and Math Enrichment

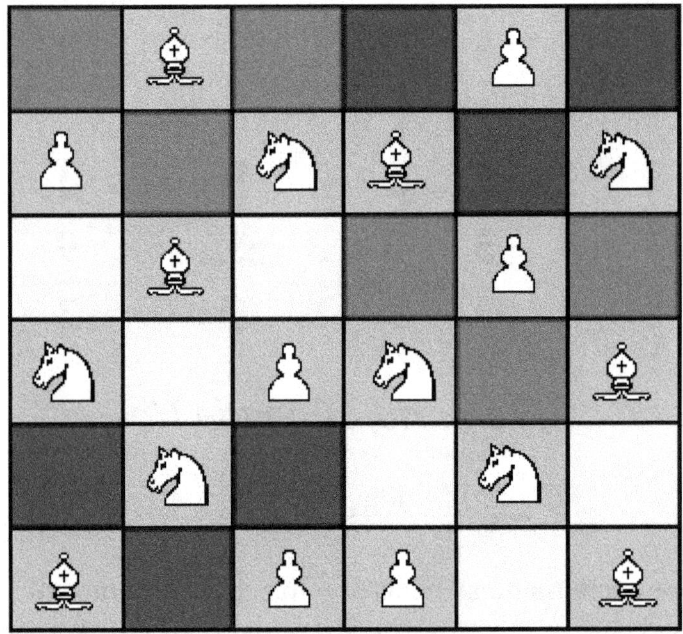

Frank Ho, Amanda Ho

mathandchess@telus.net

Ho Math Chess™ is the only international education franchise dedicated to teaching children math through its first math, chess, and puzzles integrated workbooks. These workbooks have been classroom-tested and proven to be statistically significant in raising students' math marks, improving critical thinking ability, advancing chess knowledge and increasing brainpower. **Ho Math Chess**™ is the world's leading and most reputable and largest academic math and chess franchiser worldwide.

www.homathchess.com

Student's name: ________________ Assignment date: ________________

Sample Ho Math Chess worksheets

只见棋谜不见题 劝君迷路不哭涕 数学象棋加谜题 健脑思维真神奇

Multi-grade multi-level math (多年级多功能计算题**)**

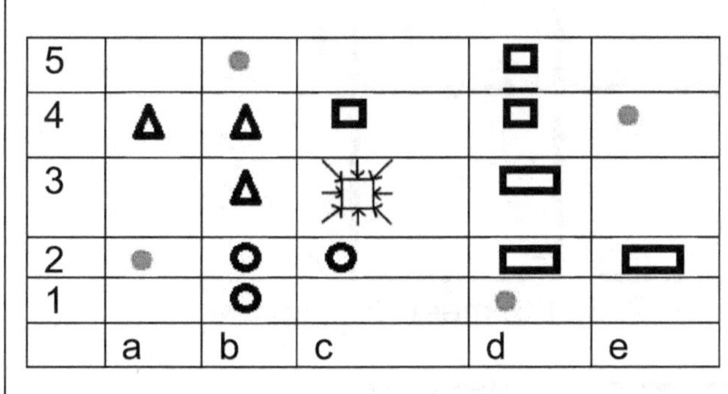

You are a chess piece located at c3.

Fraction	$\frac{1}{2}$	1	$\frac{1}{2}$	$\frac{4}{4}$	$\frac{3}{4}$
decimal	0.2	0.40	0.8	0.2	0.6
Whole	2	12	22	32	42
%	100%	200%	300%	400%	500%

Whole number

÷ = ______ ÷ ______ = ______   ÷ = ______ ÷ ______ = ______

× = ______ × ______ = ______   × = ______ × ______ = ______

Decimal

÷ = ______ ÷ ______ = ______   ÷ = ______ ÷ ______ = ______

× = ______ × ______ = ______   × = ______ × ______ = ______

Fraction of multiplication and division [Do not need to have the same measuring unit (denominator).]

× = ______ × ______ = ______   × = ______ × ______ =

÷ = ______ ÷ ______ = ______   ÷ = ______ ÷ ______ =

Fraction of addition and subtraction [Must have the same measuring unit (denominator).]

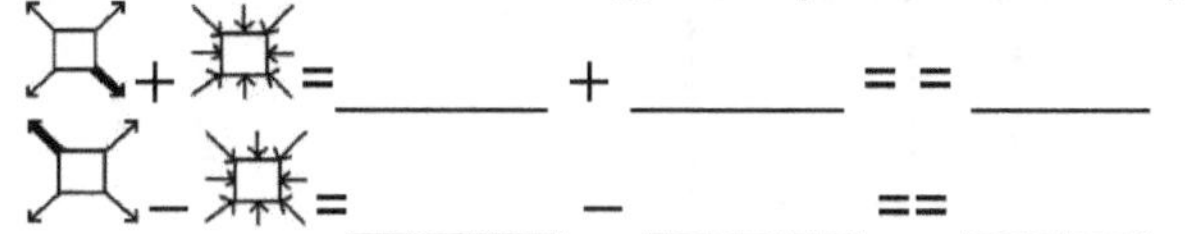

+ = ______ + ______ = ______   + = ______ + ______ = = ______

– = ____ – ____ = ______   – = ______ – ______ == ______

Percent

× = ______ × ______ = ______   × = ______ × ______ = ______

÷ = ______ ÷ ______ = ______   ÷ = ______ ÷ ______ = ______

Why buy a Ho Math Chess Franchise?

Intellectual Properties of Ho Math Chess

何数棋谜 - www.homathchess.com

Student's name: ________________ Assignment date: ____________________

Frankho ChessDoku (何数棋算独)

Use 1, 2, 3, 4 to meet Sudoku law.

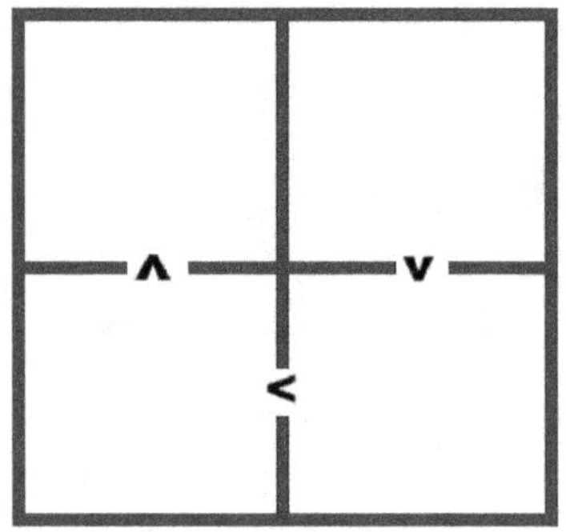

Use 1, 2, 3, 4 to meet Sudoku law.

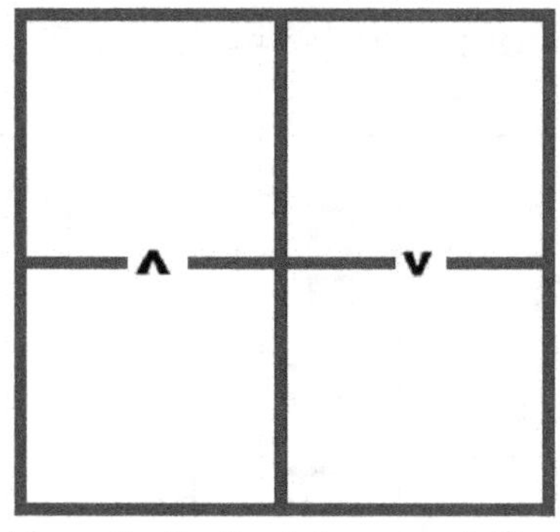

Use 1, 2, 3, 4 to meet Sudoku law.

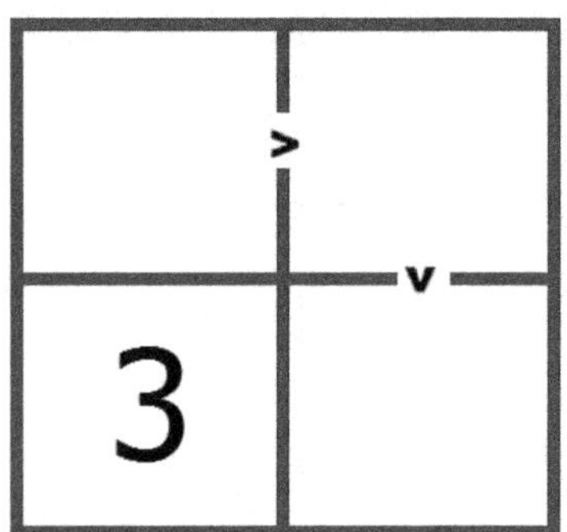

Use 1, 2, 3, 4 to meet Sudoku law.

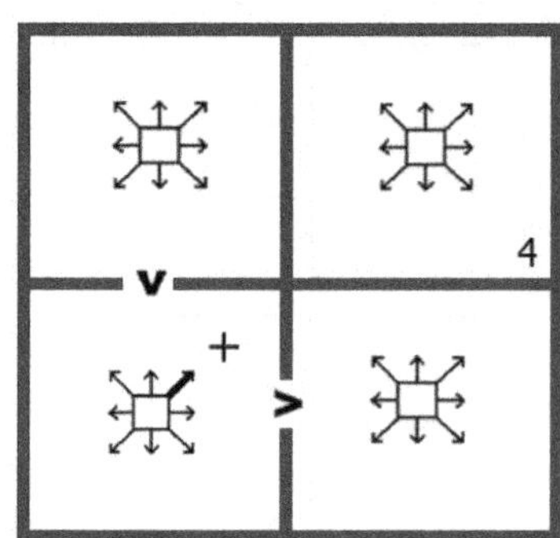

Use 1, 2, 3 to meet Sudoku law.

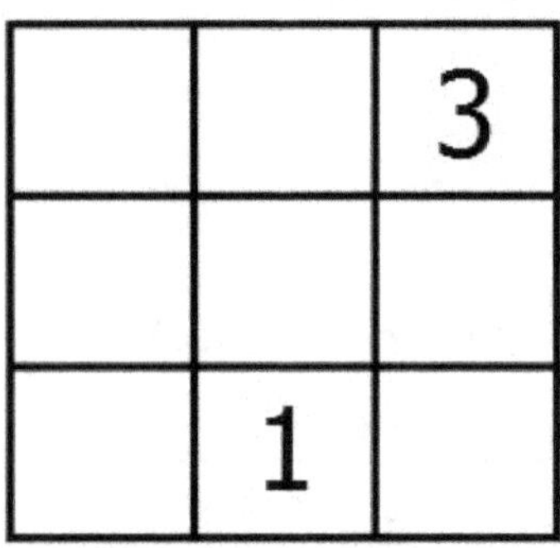

Use 1, 2, 3 to meet Sudoku law.

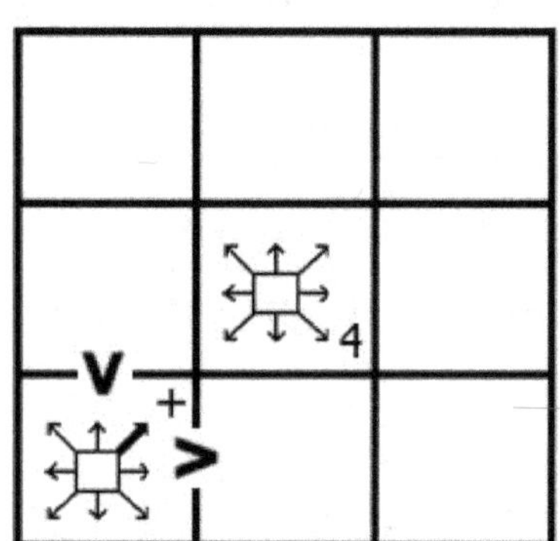

Student's name: ________________ Assignment date: ____________________

Unequal Sudoku (不等数独)

Every row and column must have only one number starting from 1 to the number of squares of each side (Sudoku), but all numbers must obey the inequality sign.

Example

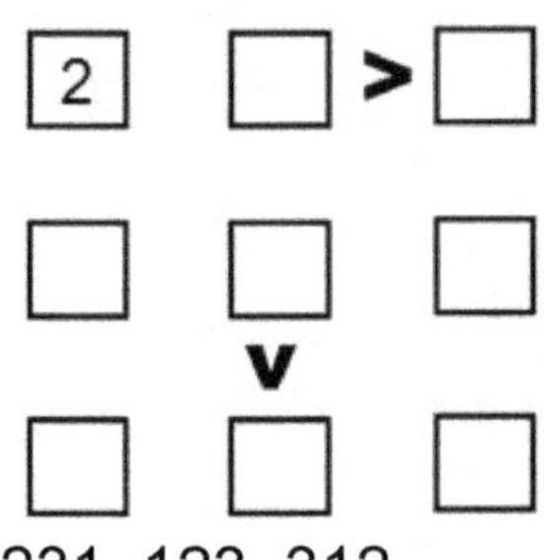

231, 123, 312

Problem

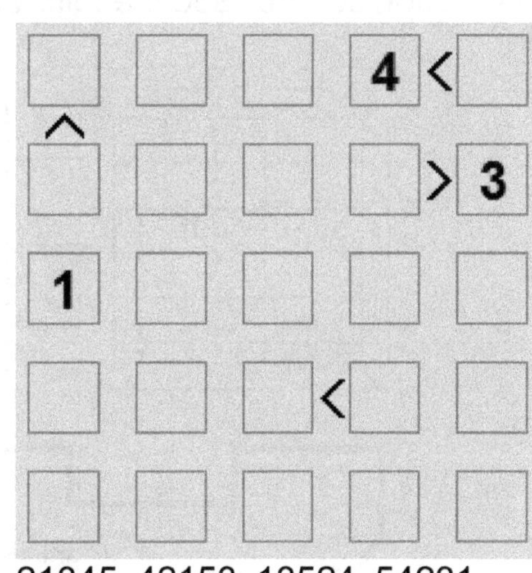

21345, 42153, 13524, 54231, 35412

Fencing (盖围墙)

Connect lines around each dot so that each number indicates how many lines, connected by four dots only, surround the dot. The connected lines must form a single loop (like one rubber band) without lines crossed to each other.

Example

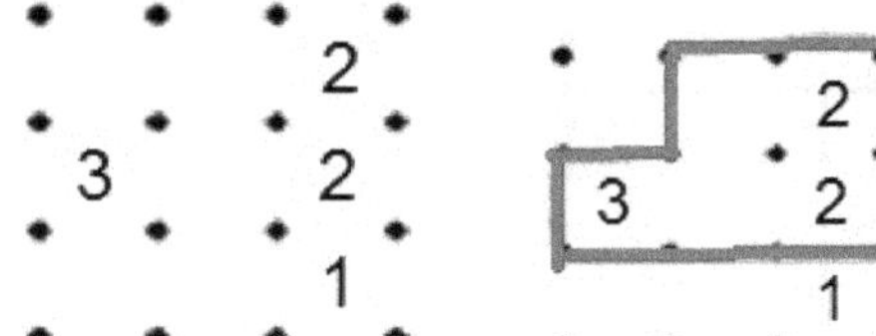

Problem

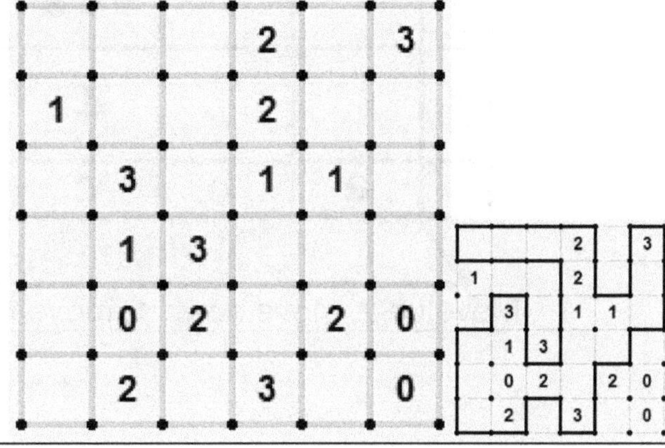

Student's name: ________________ Assignment date: ________________

Amandaho Moving Dots Puzzle ™ (移点子)

You are a rook at c3.

Move some dots in c4, c2, or d3 squares into c3 square such that the sum of dots + dots in each of rook's moves at c3 will be equal to the number shown on its destination square. See the following example.

Example

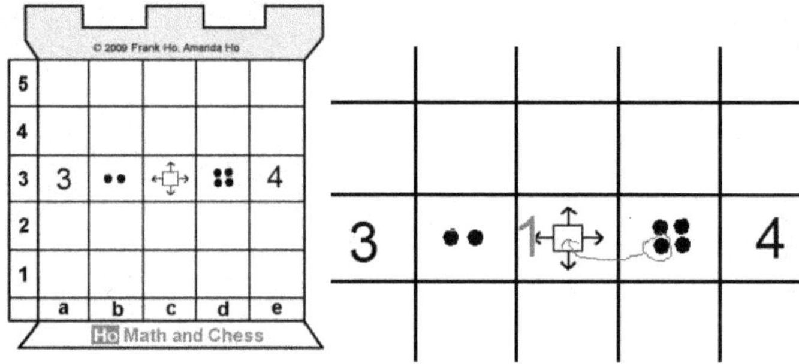

Problem

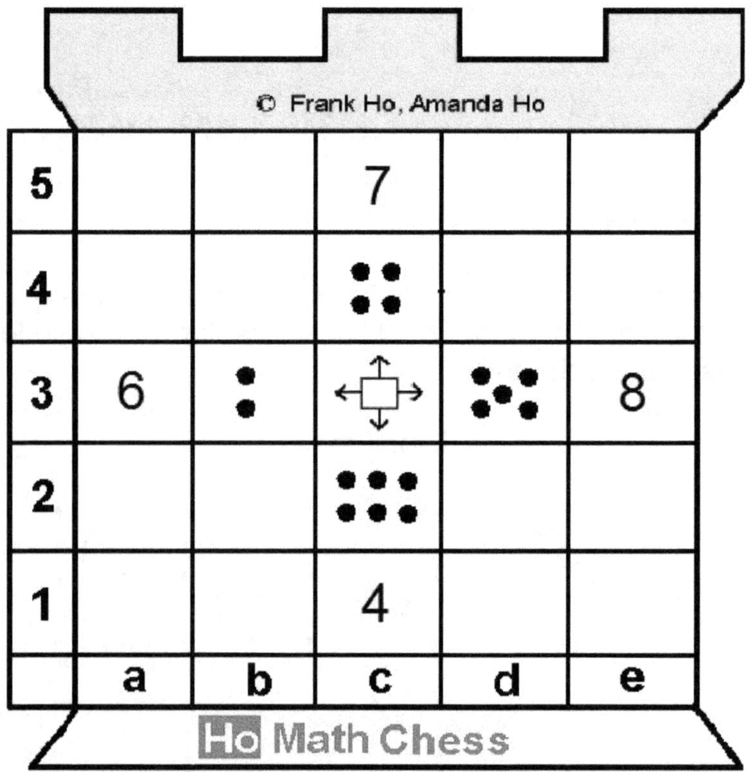

Move up 2, move down 1, move left 1. 4 in the middle.

Student's name: ________________ Assignment date: ____________________

Unequal Sudoku (不等数独)

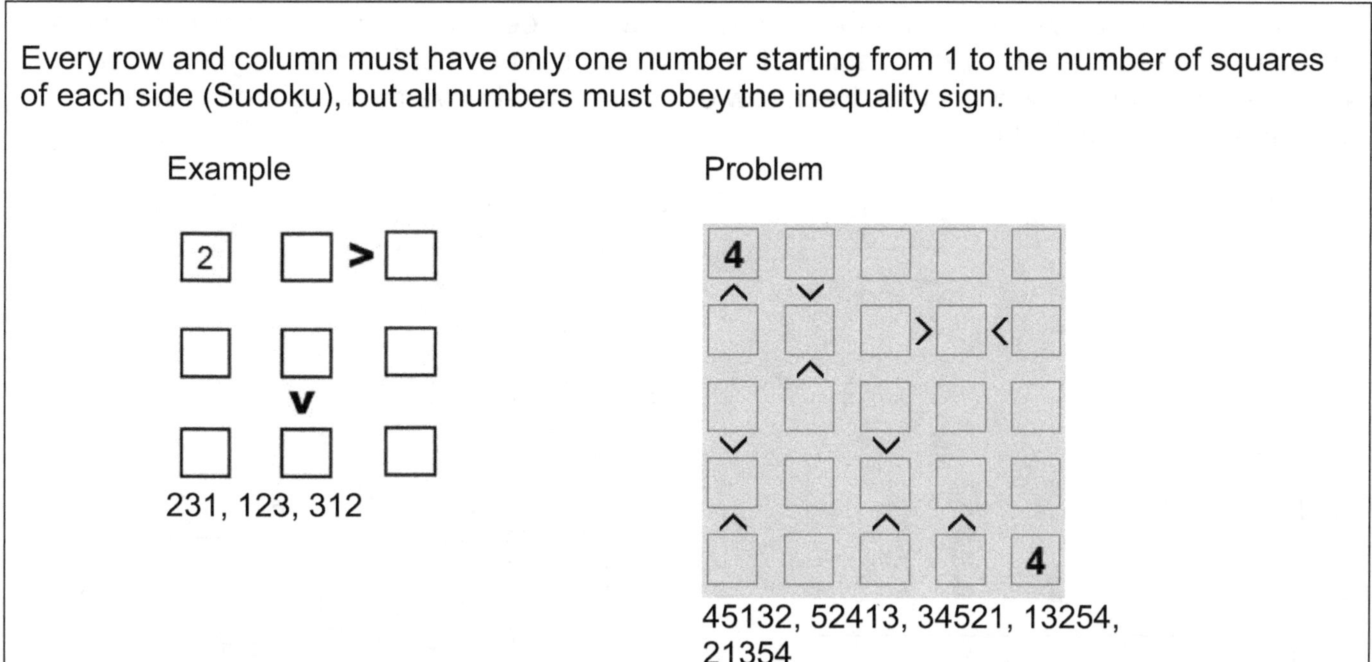

Every row and column must have only one number starting from 1 to the number of squares of each side (Sudoku), but all numbers must obey the inequality sign.

Example

231, 123, 312

Problem

45132, 52413, 34521, 13254, 21354

Fencing (盖围墙)

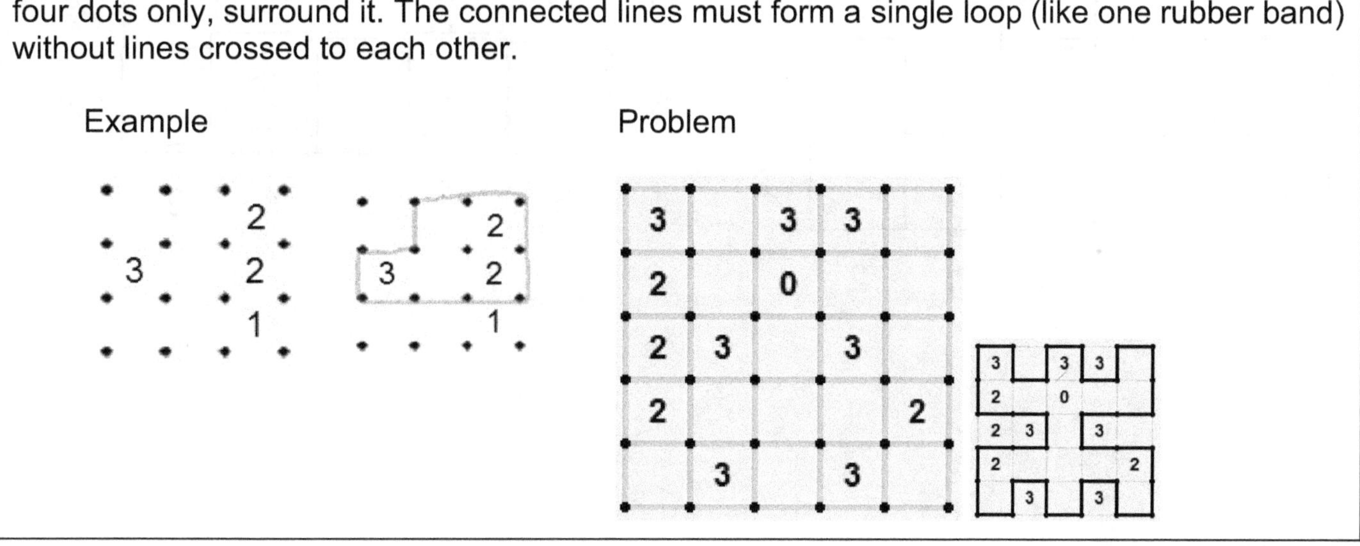

Connect lines around each dot so that each number indicates how many lines, connected by four dots only, surround it. The connected lines must form a single loop (like one rubber band) without lines crossed to each other.

Example

Problem

Why buy a Ho Math Chess Franchise?

Intellectual Properties of Ho Math Chess

何数棋谜 - www.homathchess.com

Student's name: ________________ Assignment date: ____________________

3-D Sudoku Towers (看高楼数独)

The number outside the grid shows the number of towers seen from that row or column when looking into the square from that direction. For example, the number "2" means you can see two towers from that row or column. Each row or column must have the height of a tower according to the Sudoku law.

Example

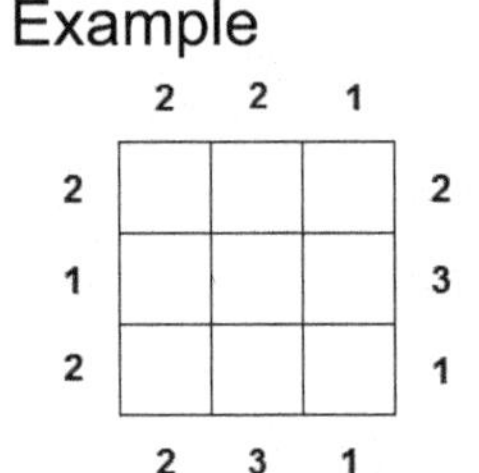

Answer

Problem

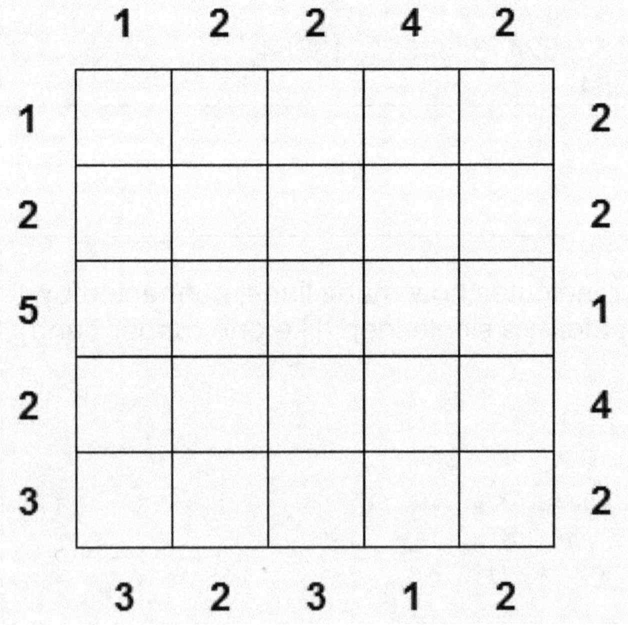

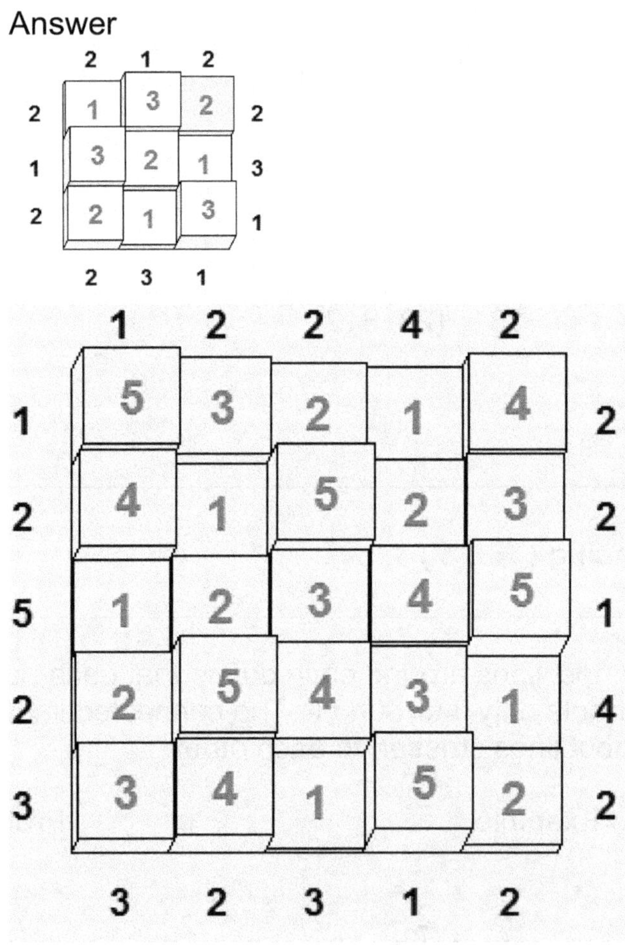

Student's name: ________________ Assignment date: ___________________

Sudoku (数独)

Each row contains only one digit from 1 to 9.
Each column contains only one digit from 1 to 9.
Each box contains only one digit from 1 to 9.

					4		1	
	1					3	5	9
		5			6			
	7	2		6	8			
			7		1			
			3	5		8	7	
			9			7		
1	9	6					8	
	3		6					

7	2	9	5	3	4	6	1	8
6	1	4	8	7	2	3	5	9
3	8	5	1	9	6	2	4	7
9	7	2	4	6	8	1	3	5
8	5	3	7	2	1	9	6	4
4	6	1	3	5	9	8	7	2
5	4	8	9	1	3	7	2	6
1	9	6	2	4	7	5	8	3
2	3	7	6	8	5	4	9	1

Student's name: ________________ Assignment date: ____________________

CalcuDoku

Each row contains only one digit from 1 to the size of the grid.
Each column contains only one digit from 1 to the size of the grid.
The digits in each block must satisfy the arithmetic operation given in the block.

Example

1−	1−	
	9×	
3+		

1− 3	1− 1	2
2	9× 3	1
3+ 1	2	3

2÷	3×		1−
		6×	
8+			2÷
	2−		

2÷ 2	3× 3	1	1− 4
4	1	6× 2	3
8+ 1	4	3	2÷ 2
3	2− 2	4	1

30×			5×	3−
6+		2−		
60×			2÷	
5×		7+		2−
	5+			

30× 2	3	5	5× 1	3− 4
6+ 4	2	2− 3	5	1
60× 3	5	1	2÷ 4	2
5× 1	4	7+ 2	3	2− 5
5	5+ 1	4	2	3

Student's name: ________________ Assignment date: ________________

What is Ho Math Chess™ Teacher Set?

The following is an image of Ho Math Chess teaching set.

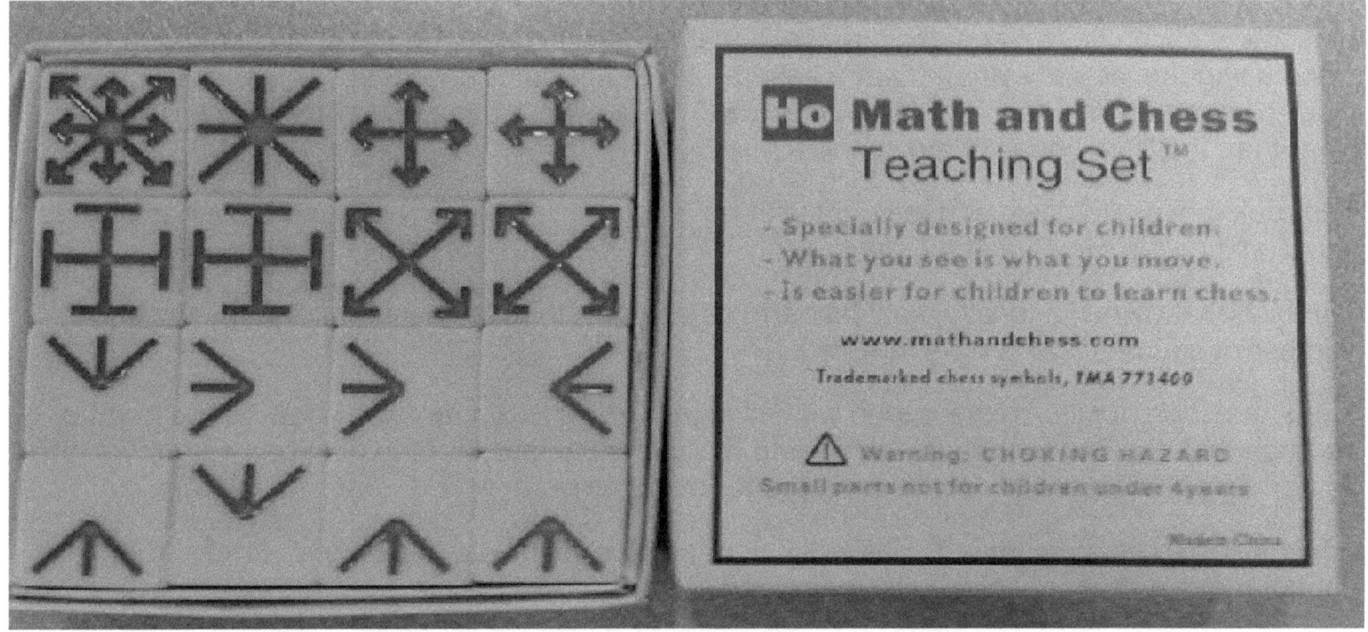

December 11, 2007, is when the world's first flat-surfaced, all-squared, uniform look chess teacher set was available to the world. This chess teacher set will become a collector's item since it is the world's first. It creates a new dimension on how math can be taught by combining chess and math through Frankho Geometry Chess Symbols (Trademark Canada TMA771400 and international copyright registered in Canada).

The theoretical background on how this set was published in Vector (Fall 2007, Volume 48, Issue 3), the British Columbia Association of Mathematics Teachers' official journal in Canada. The website for the published article is available to view at http://www.scribd.com/doc/207579234/A-New-Chess-Set-for-Teaching-Mathematical-Chess

We use this GCS in our workbooks, and it surprisingly makes children like to work on math. The fact of the matter is it also does not cause any confusion for a student who learns math. The simple reason is they are just symbols, so they do not create any cognitive dissonance or jeopardize the learning process.

One-of-its-kind and unique Ho Math Chess Teaching Set is sold worldwide and carries a reputable brand name. This Ho Math Chess teaching set is "out-of-this-world", a genius invention by using Frankho Geometry Chess Symbols

Student's name: ________________ Assignment date: ____________________

Ho Math Chess Teaching Set

How to play?

Since how each chess piece could move is marked symbolically on the surface of each chess piece using the idea of symmetry of a square and the geometric concept of the line (showing by arrows) and a line segment (with no arrows), so how to move each chess piece is much clear than the "traditional" chess set.

The rook moves up/down and left/right with no number of squares restricted, so it has arrows showing it can go/down and left/right.

The bishop moves diagonally, so it has arrows showing that its chessboard only limits the number of squares that can be reached.

The same idea applies to the queen with all eight arrows in 8 directions showing the number of squares that can be reached is only limited by its chessboard, but the king has no arrows, so its movement is limited to only one square for each move.

A pawn can move forward but captures diagonally, so its moves showing no arrows (I.e., cannot move with an unlimited number of squares).

Knight has the shape of 8 L directions but no arrows, so it can "jump" to only one directed square by a line segment in L shape.

The description of our unique chess set is as follows:

- Specially designed for children as young as 4-year old to learn chess quickly.
- Multi-function capabilities for playing traditional chess games and also blind or half-blind games to improve memory.
- Can be used to play puzzles, not just chess.
- Looking for something unique? Your wish has come true!
- A lovable, pocket-sized chess set for easy carrying. Children love it!
- What you see is what you move! So easy to play chess now!!!! It is a fantastic invention!
- World's first flat-surfaced international chess set with copyright-applied Geometry Chess Symbol marked on each chess piece!

Student's name: ________________ Assignment date: ____________________

Ho Math Chess™ Teaching Set

A wonderful chess teacher chess set

Frank Ho, founder and CEO of Ho Math Chess, invented a revolutionary chess training set especially designed and manufactured for teaching children or novice players as young as 3 or 4 years old to play chess.

This incredible chess set just plays like a standard 3-D chess set but offers additional advantages. One of them is each chess piece's moves are marked on each chess piece to make each chess piece much easier and fun for a young child to understand. It is a "what you see is what you move" chess set and teaches geometry concepts of lines, line segments, and transformations.

Besides, since it has a flat surface, children can use "blind" chess to turn each piece upside down and then flip over one by one in a standing position to have great fun. No more spills or bumps for small hands when moving pieces.

The invention of the Ho Math Chess training set has revolutionized the chess learning population profile to possibly as young as pre-kindergartners

For more details on this innovative Ho Math Chess set, please visit www.homathchess.com.

Student's name: _______________ Assignment date: ___________________

Tablero de Ajedrez Ho Math Chess™

Un fabuloso tablero de ajedrez del profesor

Frank Ho, fundador y CEO de Ho Math Chess inventó un tablero de enseñanza de ajedrez (pendiente de patente) especialmente diseñado y construido para enseñar ajedrez a niños o jugadores principiantes en edades a partir de los 3 o 4 años.

Este tablero increíble de ajedrez funciona simplemente como un juego típico en 3 dimensiones pero ofrece una ventaja adicional que consiste en que las movidas de cada pieza de ajedrez están marcadas con claridad en cada pieza para facilitar mucho el aprendizaje de las movidas y para que éste sea divertido para los niños pequeños. Se trata de un tablero en la forma de "lo que ves es lo que mueves" y que además enseña conceptos de geometría sobre líneas y segmentos de líneas, así como transformaciones.

Además, dado que tiene una superficie plana, de manera que pueda ser usado por los niños para jugar un ajedrez "ciego", es decir para mover todas las piezas de arriba hacia abajo y luego girarlas para divertirse mucho. Al mover las piezas ya no veremos más caídas o tropiezos con las manos pequeñas.

La invención del tablero de enseñanza Ho Math Chess ha revolucionado el perfil de la población en edad de aprender ajedrez probablemente hasta la edad de preescolar.

Para mayor detalle sobre este tablero innovador de Ho Math Chess, visite www.mathandchess.com.

Student's name: ________________ Assignment date: ____________________

To see the table of contents, please click on the BCAMT logo.

The Official Journal of the British Columbia Association of Mathematics Teachers.

Student's name: ________________ Assignment date: ____________________

FRANK HO

Frank Ho is a teacher at the Ho Math and Chess Learning Centre (http://www.mathandchess.com).

A New Chess Set for Teaching Mathematical Chess

If one thinks that chess consists of warriors or commanders and kings and queens battling in the field, then this notion does not really address the reason why chess pieces move in pattern-like directions. For example, a rook moves up and down or left and right and a bishop moves diagonally. Is chess a reflection of ancient war or is it an invention based on a mathematical principle? The author believes that chess was invented by using the concept of geometric symmetry. This conjecture is based on analyzing the moves of each chess piece and I thus conclude that chess was created from a mathematical point of view.

For a fair game, the positions of the chess pieces and the layout of the chessboard must be symmetric. Perhaps it is not coincidental that the playing field of chess is all about squares and that the Chinese character for "rice field" is also a 2 by 2 square. Given a fixed perimeter, the square has the largest area. The chessboard is an 8 by 8 tessellation of 64 squares. There are 4 lines of symmetry in each square and these 4 lines constitute the moves of the rook, queen, king, pawn and bishop. It makes sense that each chess piece moves along these lines.

How Chess Moves Originated

To play a symmetric game, the smallest board required is 5 by 5. I believe that the possible moves of each chess piece are originally intended to be 360 degrees of circular movement. For example, take a look at a 5 by 5 chessboard. (See figure 1.) If a chess piece is placed at c3, how many ways can this chess piece reach out to form the shape of a circle? In addition to a circle, a square shape could also be formed, depending on how the points are connected.

The first "easy" way would be to move down or up and left or right, from c3 so as to reach the limit of a circle. In this way, the moves of the rook are born. Its motion is called a translation or a slide in geometry. If you connect the 4 out-reached points with 4 straight lines, the shape is actually a square; however with contours it then forms a circle.

The second way of moving to form a circle is to move in the directions of two main diagonals. A circle is thus produced and the moves of a bishop are born. Arguably, the four points also make a square shape. This motion from c3 to each of the 4 diagonal points is also a double-slide. The bishop can move in 360 degrees.

Combining the above two ways of a rook and bishop moving, we have the most powerful of all chess pieces: the queen. This is the birth of a queen's move. Finally, the king follows the moves of a queen, but can move only one square per turn.

In a 5 by 5 chessboard, we notice that all chess squares on each of 4 sides are covered by the moves of a rook and bishop except a2, a4, b1, b5, d1, d5, e2, and e4. So from an attacking or defending point of view, this is a problem: there are 8 squares that are not covered. This is the reason for the birth of another chess

Student's name: ________________ Assignment date: ____________________

The Effect of Math and Chess Integrated Instruction on Math Scores

John BUKY, Education Consultant
Frank HO, Canada certified math teacher

Frank Ho is a Canadian certified math teacher with a Master of Science degree from Brigham Young University, Utah, USA. He is now working as a private math tutor in Vancouver, Canada.

John Buky is a USA certified teacher and has two master's degrees: an M.A. in Education from Olivet Nazarene University and a Master's in Instructional Technology from Grand Canyon University. John is now working as a private educational consultant in Chicago, Illinois, USA.

This article is published by The Chess Academy, Chicago, USA, June 2008

Research studies have shown that chess can be used as an effective game-based teaching method. However, all past studies used chess as a separate instructional tool. There were no math contents in chess instruction provided, and there was no math and chess integrated workbook used. This study examined pupils' math scores when genuinely integrated math and chess workbooks were used as an instructional practice workbook. The results show that integrated math and chess workbook significantly increased pupils' math scores between pre-tests and post-tests from grade 1 to grade 8 pupils.

Key Words: math and chess; math and chess instruction, math and chess integrated workbook; math and chess integrated workbook; mathematics scores of the students

Introduction

Research papers have demonstrated that chess instruction improves analytical reasoning, problem-solving skills, and academic achievement (Chrisiaen & Verholfstadt (1978); Frank & D'Hondt (1979); Smith & Cage (2000)). Research conducted by Gaudreau (1992) shows no significant differences among the groups on necessary calculations. These research studies point to the direction that chess has a strong effect on improving children's cognitive ability than their arithmetic computation ability. By teaching math and chess as two separate subjects, children do not have opportunities to work on basic arithmetic operations using acquired chess knowledge; this may explain why playing chess may not statistically significantly improve children's basic arithmetic computation ability.

How to maximize the benefits of chess instruction in such a way that not only chess benefits children's cognitive development but also their computation ability? All the past chess instruction research studies have used chess instruction as an independent teaching tool, and it is not truly integrated with math instruction. The author Frank Ho created a math and chess integrated workbook. The theoretical basis of how math and chess are integrated has been published by Ho (2006). We believe that with the creation of truly integrated math and chess workbooks, pupils will increase their computation ability by working on these math and chess integrated workbooks. This is particularly important for those children who have no interest in playing chess. However, they could still benefit from chess instruction by working on math and chess integrated workbooks.

Student's name: ________________ Assignment date: ____________________

A simple example below demonstrates an idea of a math and chess integrated worksheet: How many points altogether is the square ☒ or ☒ being attacked?

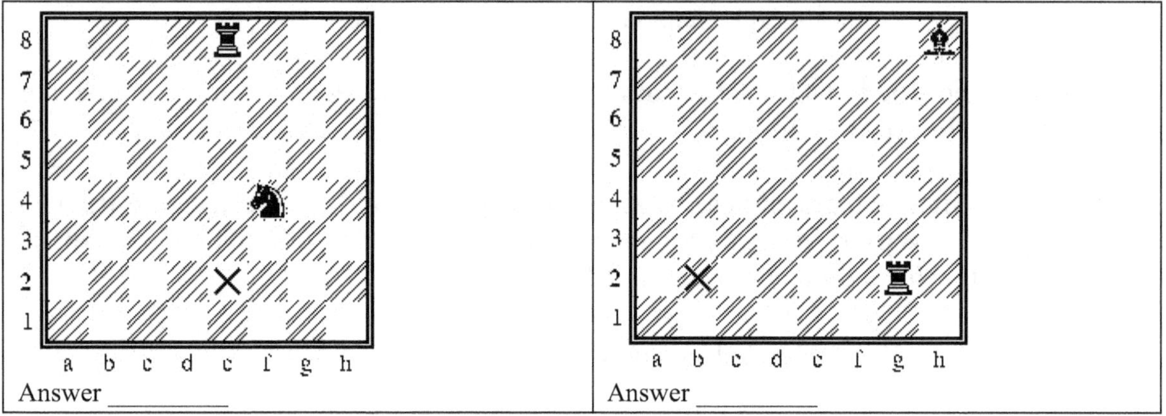

No research has been done before on the effects of using math and chess integrated workbook. This study will compare the effect of pupils' math computation ability before using the math and chess integrated workbook and after using it to see if there is a significant difference.

Method

In grade 1 to grade 8, one hundred and nineteen pupils from five public elementary schools in Chicago, Illinois, USA, participated in the after-school program for 120 minutes, twice a week, for a total of 60 hours of instruction. None of the students has possessed any substantial knowledge of chess. The study began by administering pre-tests in the first week of this study at the beginning of the program on 10/23/06, and a post-test was conducted at the end of the program on 3/28/07. Tests of TONF (The Compass Learning Explorer Online Diagnostic Tool) were used for the pre-test and post-test. The Compass Learning Explorer Assessment meets the requirements as a handy and reliable criterion-referenced assessment tool.) were given to all pupils for both tests. Each lesson consisted of lecturing, practice on math and chess integrated worksheets and chess playing.

Results

The paired t-test was used to analyze the data. This study's results show significantly different in their math scores for all grade 1 to grade 8 pupils between pre-test and post-test at the level of p is less than 0.01.

Group	Group One	Group Two
Mean	36.46	55.45
SD	15.82	19.37
SEM	1.45	1.78
N	119	119

t = 12.8729

Student's name: ________________ Assignment date: ___________________

Discussion

This study demonstrates that a genuinely integrated math and chess workbook can significantly improve pupils' math scores. Our observations show that the effect of using a truly integrated math and chess workbook also provides mental entertainment and thought by pupils as more fun than traditional computation practices. Pupils could sit longer when working on math and chess integrated workbooks than working on traditional computation worksheets.

This research is particularly interesting for children who do not have a high interest in playing chess since math and chess integrated workbook involves visualization, analyzing, spatial relation, and data processing. These types of problems provide high-order cognitive skills. Without spending substantial time playing chess, we believe that children can get similar benefits of playing chess on cognitive effects by working on math and chess integrated workbooks. This may require further study.

Why do children like to work on math and chess integrated workbooks than on the traditional computation worksheets? Math and chess integrated work have visual images, chess symbols, directions, spatial relation, and tables; all these are stimuli to kids and keep their interests high while working on computation problems. This also gives children ample opportunities to think visually. Most of the time, the computation questions themselves are not written for children to work on immediately but for children to "create" themselves. These questions have to be actually "mapped" out by following directions, and children love them. Children learn best while having fun.

References

Chrisiaen and Verholfstadt, (1978) "Chess and cognitive development", Nederlandse Tydschrift voor de Psychology en haar Grensbebieden 36, 561-582.

Frank and D'Hondt, (1979) "Aptitudes and learning chess in Zaire", Psychopathologie Africane, 15, 81-98.

Gaudreau (1992), "Etude Comparative sur les Apprentissages en Mathematiques 5e Annee", June manuscript

Ho Frank (2006), "Enriching math using chess," ***Journal of the British Columbia Association of Mathematics Teachers***, British Columbia, Canada, Vector, Volume 47, Issue 2.

Smith & Cage (2000), "The effects of chess instruction on the mathematics achievement of Southern, rural, Black secondary students," Research in the Schools, 7, 19-26

Student's name: ________________ Assignment date: ____________________

Ho Math Chess Method

The setup of a chessboard using real flat chess pieces is as follows:

The setup of a chessboard using the geometric chess symbols is as follows:

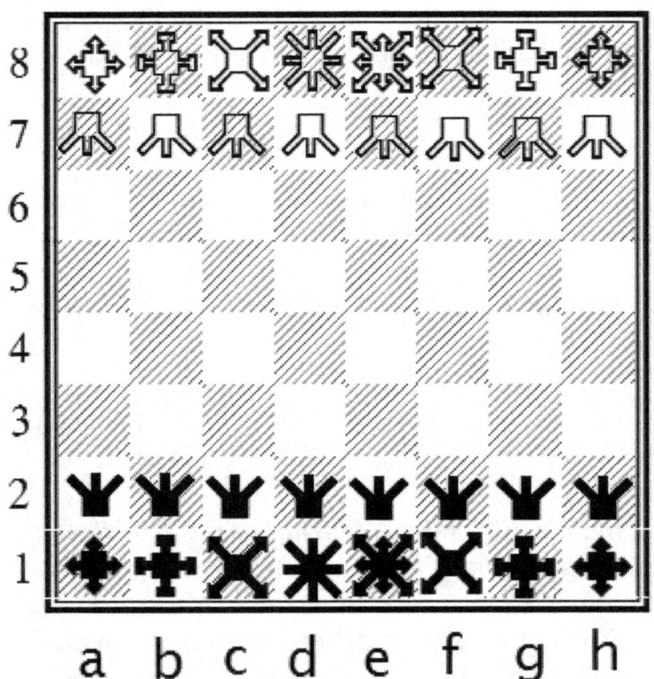

Student's name: ________________ Assignment date: ___________________

The Ho math Chess method, developed by a Canadian math teacher Frank Ho [1], is a mathematics teaching method by integrating chess and puzzles into the mathematics curriculum.

History

Frank Ho taught chess to his son when he was five years old[2]. His son was the Canadian junior chess champion at age 14[3] and later was awarded FIDE chess master [4].

In 1995, Ho was intrigued by the relationship between chess and math and took the initiative to write math and chess integrated workbook entitled Mathematical Chess Puzzles for Juniors (ISBN 0-9683967-0-4, library catalogue number QA95.H6 1998 j793.7'4). The theoretical basis of Ho's math and chess teaching method was published in the Official Journal of the British Columbia Association of Mathematics Teachers in Canada [5].

The Program

Ho believes children learn best while having fun, so he integrated chess and puzzles into math teaching. This way, children learn math, chess, and puzzles simultaneously, and children also have the opportunities to work on math, chess, and puzzles integrated worksheets. Ho invented Frankho ChessMaze and Frankho ChessDoku [6]. He also invented a chess teaching set to learn chess with ease as young as 4 years old. The concept of this chess set is published in [5].

Student's name: ________________ Assignment date: ____________________

Ho Math Chess learning centre

In 1995, Ho opened his first Ho Math Chess learning centre in Vancouver, Canada, and after 2004, Ho Math Chess began to open around the world [7].

References

1. British Columbia Teacher Certificate No. 2008/0584.
2. How My Son Became a Chess Master, http://searchwarp.com/swa285033.htm
3. http://www.chesscanada.info/forum/showthread.php?t=552
4. 2003, he was awarded FIDE chess master. http://web.ncf.ca/bw998/Titles.html
5. Vector, Fall 2007, Volume 48, Issue 3
6. Canadian Intellectual Property Office copyright number certificate registration number 1069744
7. Entrepreneur, http://www.entrepreneur.com/franchises/homathandchesslearningcentre/321831-0.html

Student's name: ________________ Assignment date: ____________________

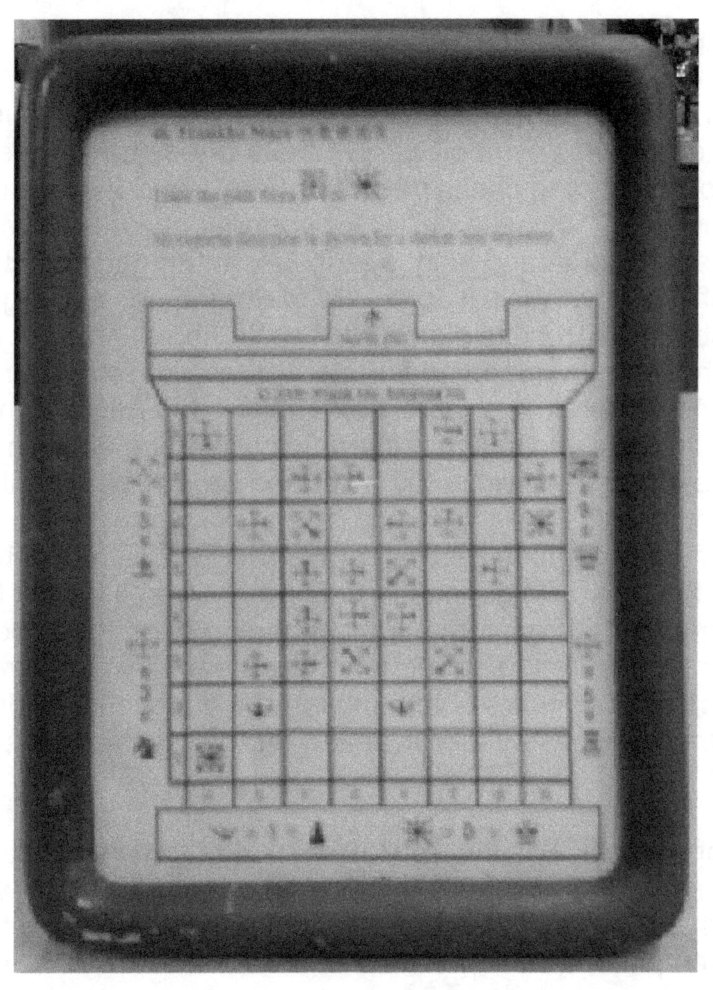

Student's name: ______________ Assignment date: ________________

Symbolic Chess Language

Frank Ho, the Ho Math Chess Learning Center founder, headquartered in Vancouver, Canada, invented a set of geometric chess symbols or Symbolic Chess Language (SCL). Each chess symbol represents a corresponding chess piece (Figures 1 and 2). This new set of chess symbols not only makes teaching chess easier for younger children as young as four years old, but it also serves as a set of command language to link arithmetic and chess. The teaching idea of using this set of chess symbols is to create math and chess integrated problems or any variations of future problems as results of using these symbols, This set of chess symbols and their teaching method have been approved for intellectual property international copyright protection. Problems shown herein are merely exemplary and may be changed to suit different types of problems. Accordingly, the inventor intends to embrace all such alternatives, modifications, and variations as fall within the spirit and broad scope of this invention.

As far as the training of playing chess itself is concerned, SCL symbols have many advantages over the regular chess fonts or 3 D figurines in training children's critical thinking skills when integrating chess into math. The transformation concept in geometric chess symbols is self-explanatory. It is easier for children to understand when each symbol direction is pointed by an arrow representing each chess piece's actual movement direction. Children get hands-on experience in moving those pieces by merely following the directions displayed on each chess piece.

The other advantage of using SCL is that problems related to spatial relation, pattern, and shapes are created using geometric chess symbols to provide children with opportunities in learning important math concepts in patterns, sequence, symmetry and transformation related math problems. For example, a typical problem might involve how a 3D object such as a chess piece (♛) is transformed into a symbol () and then a symbol ()is translated into a number (9), and finally, a numerical value is produced as an answer.

A mathematical symbol language using the SCL set can be developed to create an array of innovative arithmetic problems. These geometric chess symbols themselves represent the moving direction of chess figurines. For example, a black rook is represented by this symbol  and a highlighted arrow, such as  indicating its direction of movement towards the right. In this rook case, the symbol not only can represent the chess piece itself, but it also has another attribute, which has the four directions (up, down, left and right) of moving. The directions can be one way, two ways, three ways, or four ways, so altogether, there could be 13 ways of combinations of moving directions. A simple rooks' move problem could become a very challenging problem when combined with arithmetic computation problems.

Student's name: ________________ Assignment date: ____________________

The effect is children feel thrilled and are more willing to work on chess and math combined problem since each problem requires children's creativity to create the questions by following a puzzle-like mini question and the requirement of having children to write the questions reinforces the task of memorizing the basics facts of addition, subtraction, or multiplication without causing stress on children.

Ho Math Chess ™ believes the invention of this SCL has brought integrated math and chess teaching to a new horizon. We are very proud to be the leader in the continued research of math and chess integrated teaching.

Figure 1 Geometric chess symbols for black pieces

Points	**1**	**5**	**3**	**3**	**0**	**9**
Symbols of traditional chess pieces	♟	♜	♞	♝	♚	♛
English name	**Pawn**	**Rook**	**Knight**	**Bishop**	**King**	**Queen**
Symbols of flat chess pieces	♙	♖	♘	♗	♔	♕

Figure 2 Geometric chess symbols for white pieces

Points	**1**	**5**	**3**	**3**	**0**	**9**
Symbols of traditional chess pieces	♙	♖	♘	♗	♔	♕
English name	**Pawn**	**Rook**	**Knight**	**Bishop**	**King**	**Queen**
Symbols of flat chess pieces	♙	♖	♘	♗	♔	♕

Student's name: ________________ Assignment date: ____________________

Set up of SCL symbols

The set up of the geometric chess symbols is as follows:

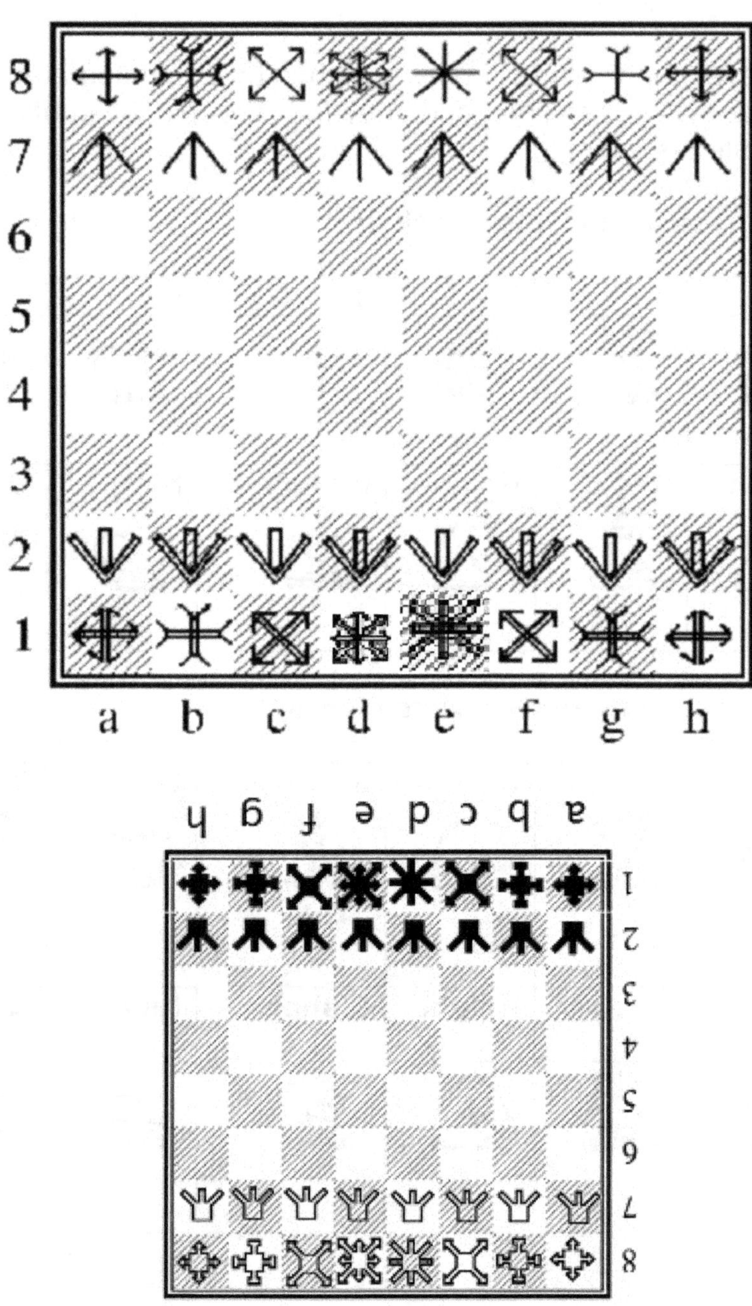

Student's name: ________________ Assignment date: ____________________

Characteristics of SCL

Each symbol carries five attributes: value, direction, colour, background colour, and size.

Value

Each symbol has its value. For example, the following gives corresponding points for each symbol for Black.

Points	1	5	3	3	0	9
Geometric chess symbols						

The static value of each symbol turns symbols into numerical values in a meaningful way. The transition and training of understanding abstract concept into forming a concrete numeric value are considered a milestone in a child's mental development. A simple example such as

+ = 3 + 3 = 6.

Direction

SCL also serves as a set of command language when its direction on the symbols is highlighted. For example, the following symbol instructs a queen to move in the northern direction.

When the above symbol is used in combination with a chessboard, a result can be achieved by asking the child to perform an arithmetic calculation.

The following demonstrates the concept described above.

Student's name: ________________ Assignment date: ____________________

3	4	5	7	9
1				6
4		♕		4
6				5
7	6	8	4	3

Calculating follows directions

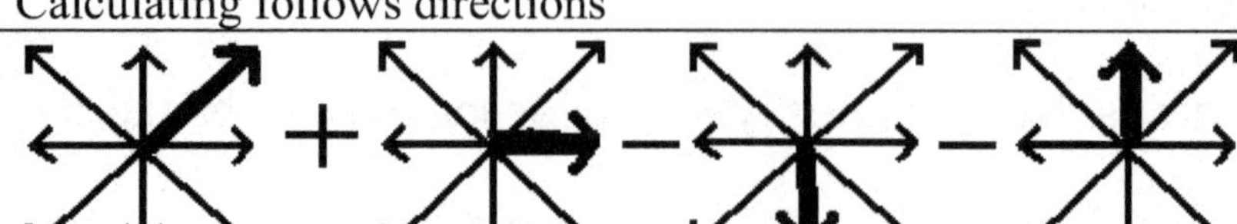

Calculate the result of the numbered squares as directed above ________.

9 + 4 – 8 – 5 = 13 – 8 – 5 = 5 – 5 = 0

More complicated problems can be created by combining the direction with coordinates specified. The following example demonstrates the idea described.

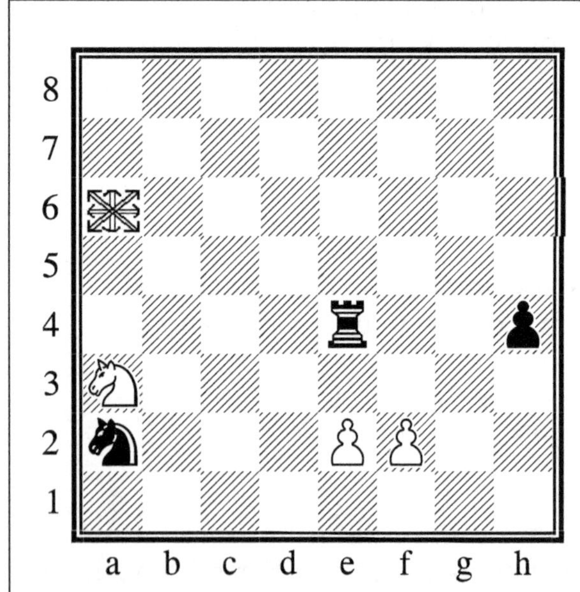

Move ♜ according to the instructions below.

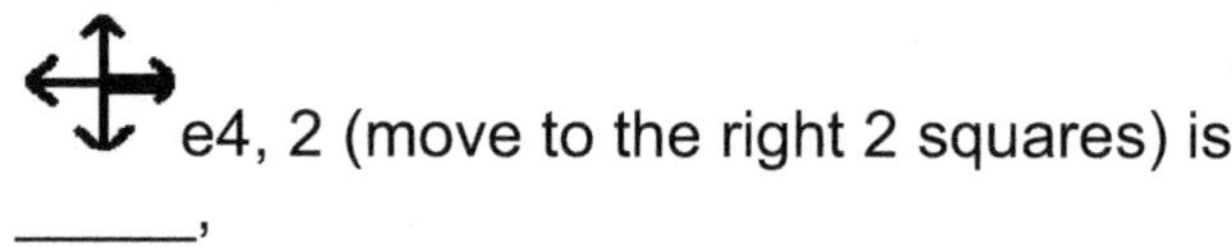

e4, 2 (move to the right 2 squares) is ______,

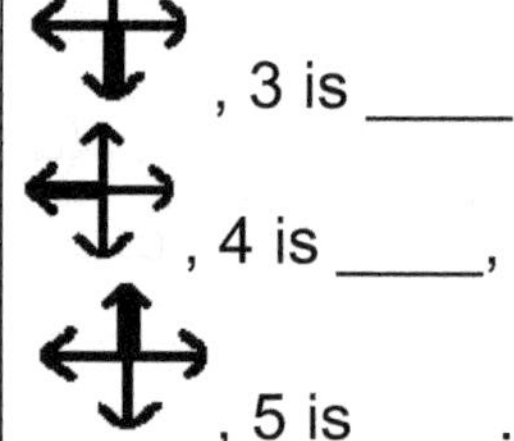

, 3 is _____,

, 4 is _____,

, 5 is_____.

Student's name: ________________ Assignment date: ____________________

A reverse problem of the above example requires a child to think backwards. The following example demonstrates the idea.

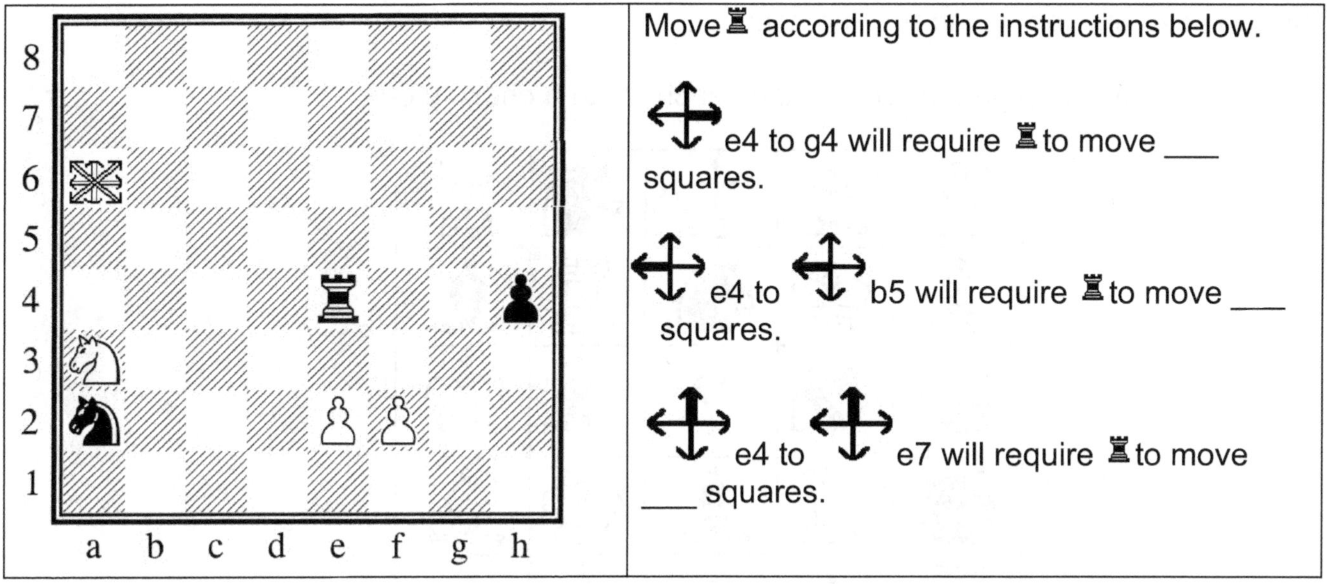

Student's name: ________________ Assignment date: ____________________

Color, Background Color and different sizes

The black and white colours of chess symbols and their black or white squares and their different font sizes present the possibilities of creating a variety of pattern problems. The following example demonstrates the concept.

Observe the following pattern and replace each? with a chess piece.

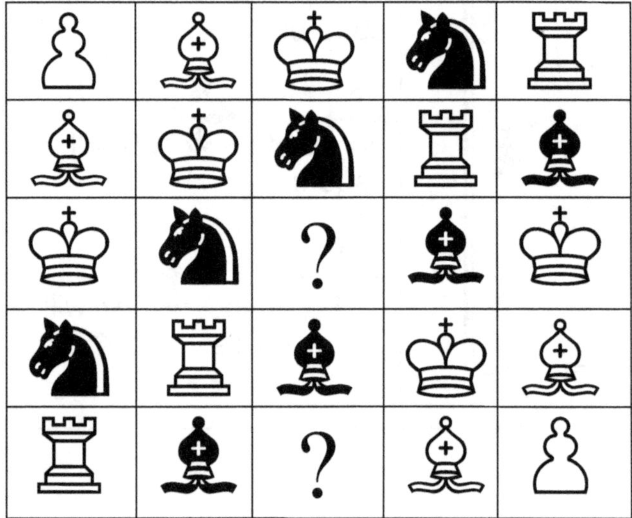

The following demonstrates how some repetitive and boring basic computation questions can be turned into puzzle-like questions using SCL and require children's creativity to figure out each question. The requirement of having children write out each question reinforces the learning outcome.

Student's name: ________________ Assignment date: __________________

3	4	5	6	2	3	4	5
7	8	9	10	6	7	8	9
11	12	13	14	10	11	12	13
15	16	17	18	14	15	16	17
4	5	6	7	5	6	7	8
8	9	10	11	9	10	11	12
12	13	14	15	13	14	15	16
16	17	18	19	17	17	18	19

Product of ✕			of		of		= ____ × ____ = ____
Product of ✕			of		of		= ____ × ____ = ____
Product of ✥	of		of		of		= ____ × ____ = ____
Product of ✥	of		of		of		= ____ × ____ = ____
Product of ✥	of		of		of		= ____ × ____ = ____
Product of ✥	of		of		of		= ____ × ____ = ____

Student's name: ________________ Assignment date: ____________________

These are just some sample questions that use Ho Math Chess' invention of SCL (Symbolic Chess Language). These puzzle-like questions allow children to create the specific questions themselves by following SCL commands using image processing with the comparison, spatial relation, logic, and interactions etc.

27 39 48 29 3 58 78 18 17	12 21 31 41 3 52 62 71 82	13 19 15 18 3 14 17 12 16	12 21 18 15 3 24 6 9 27
 39 + 3 = 42	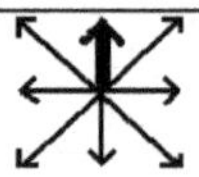 21 – 3 = 18	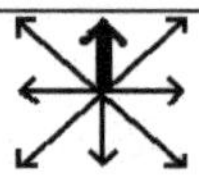 19 × 3 = 57	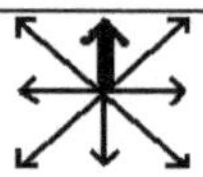 21 ÷ 3 = 7
 ___ + ___ = ___51	 ___ – ___ = ___9	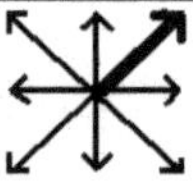 ___ × ___ = ___45	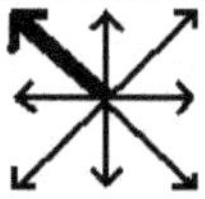 ___ ÷ ___ = ___4
 ___ + ___ = ___	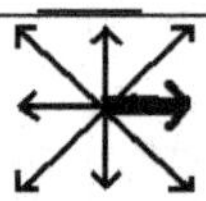 ___ – ___ = ___	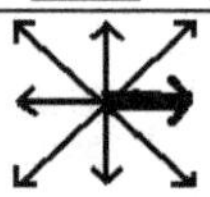 ___ × ___ = ___	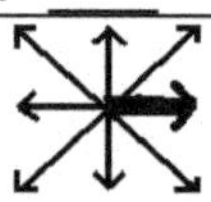 ___ ÷ ___ = ___
 ___ + ___ = ___	 ___ – ___ = ___	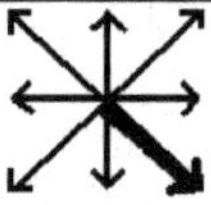 ___ × ___ = ___	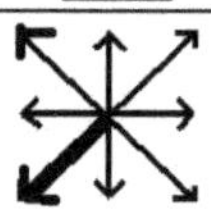 ___ ÷ ___ = ___
 ___ + ___ = ___	 ___ – ___ = ___	 ___ × ___ = ___	 ___ ÷ ___ = ___
 ___ + ___ = ___	 ___ – ___ = ___	 ___ × ___ = ___	 ___ ÷ ___ = ___
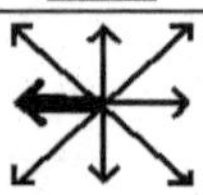 ___ + ___ = ___	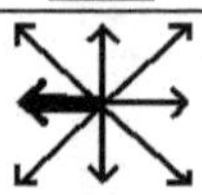 ___ – ___ = ___	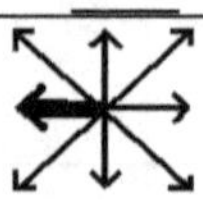 ___ × ___ = ___	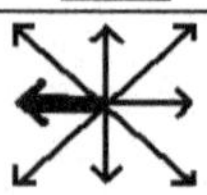 ___ ÷ ___ = ___
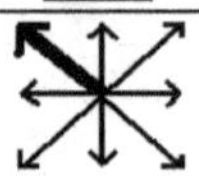 ___ + ___ = ___	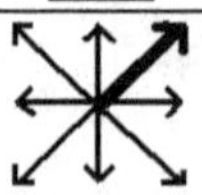 ___ – ___ = ___	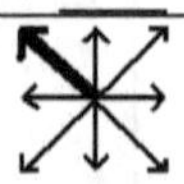 ___ × ___ = ___	 ___ ÷ ___ = ___

Student's name: ________________ Assignment date: ____________________

Multiplication using function concept and spatial relation

3	4	5	6	2	3	4	5
7	8	9	10	6	7	8	9
11	12	13	14	10	11	12	13
15	16	17	18	14	15	16	17
4	5	6	7	5	6	7	8
8	9	10	11	9	10	11	12
12	13	14	15	13	14	15	16
16	17	18	19	17	17	18	19

Product of ╳			of		of		= ____ × ____ = ____ 10×15=150
Product of ╳			of		of		= ____ × ____ = ____ 14×11=151
Product of ✥	of		of		of		= ____ × ____ = ____ 10×11=110
Product of ✥	of		of		of		= ____ × ____ = ____ 14×15=210
Product of ✥	of		of		of		= ____ × ____ = ____ 11×15=165
Product of ✥	of		of		of		= ____ × ____ = ____ 14×10=140

Student's name: ________________ Assignment date: ____________________

Multiplication using function concept and spatial relation

3	4	5	6	2	3	4	5
7	8	9	10	6	7	8	9
11	12	13	14	10	11	12	13
15	16	17	18	14	15	16	17
4	5	6	7	5	6	7	8
8	9	10	11	9	10	11	12
12	13	14	15	13	14	15	16
16	17	18	19	17	17	18	19

Product of ✕			of		of		= ______ ✕ ______ = ______ 2 x 7=14
Product of ✕			of		of		= ______ ✕ ______ = ______ 6 x 3=18
Product of ✛	of		of		of		= ______ ✕ ______ = ______ 2 x 3=6
Product of ✛	of		of		of		= ______ ✕ ______ = ______ 6 x 7=42
Product of ✛	of		of		of		= ______ ✕ ______ = ______ 3 =21
Product of ✛	of		of		of		= ______ ✕ ______ = ______ 2 x 6=12

Student's name: ________________ Assignment date: ____________________

Addition

e	11	12	13	14	15
d	16	17	18	19	20
c	21	22	2	23	24
b	25	26	27	28	29
a	30	31	32	33	34
	1	**2**	**3**	**4**	**5**

Move by one square at a time from the original square.

For example,

No □→□→□

The original square is at c3

C3 + ✥ + ✥ = _____ + _____ + _____ = _______

2 + 23 +23 = 48

The original square is at c3

C3 + ✥ + ✥ = _____ + _____ + _____ = _______

2 + 18 +18 = 38

The original square is at c3

C3 + ✥ + ✥ = _____ + _____ + _____ = _______

2 + 22 +22 = 46

The original square is at c3

C3 + ✥ + ✥ = _____ + _____ + _____ = _______

2 + 27 +27 = 56

Why buy a Ho Math Chess Franchise?

Intellectual Properties of Ho Math Chess

何数棋谜 - www.homathchess.com

Student's name: ________________ Assignment date: ____________________

Rook path

How many ways can ♜ travel to ✥ bypassing each square only once? Mark the path by lines.

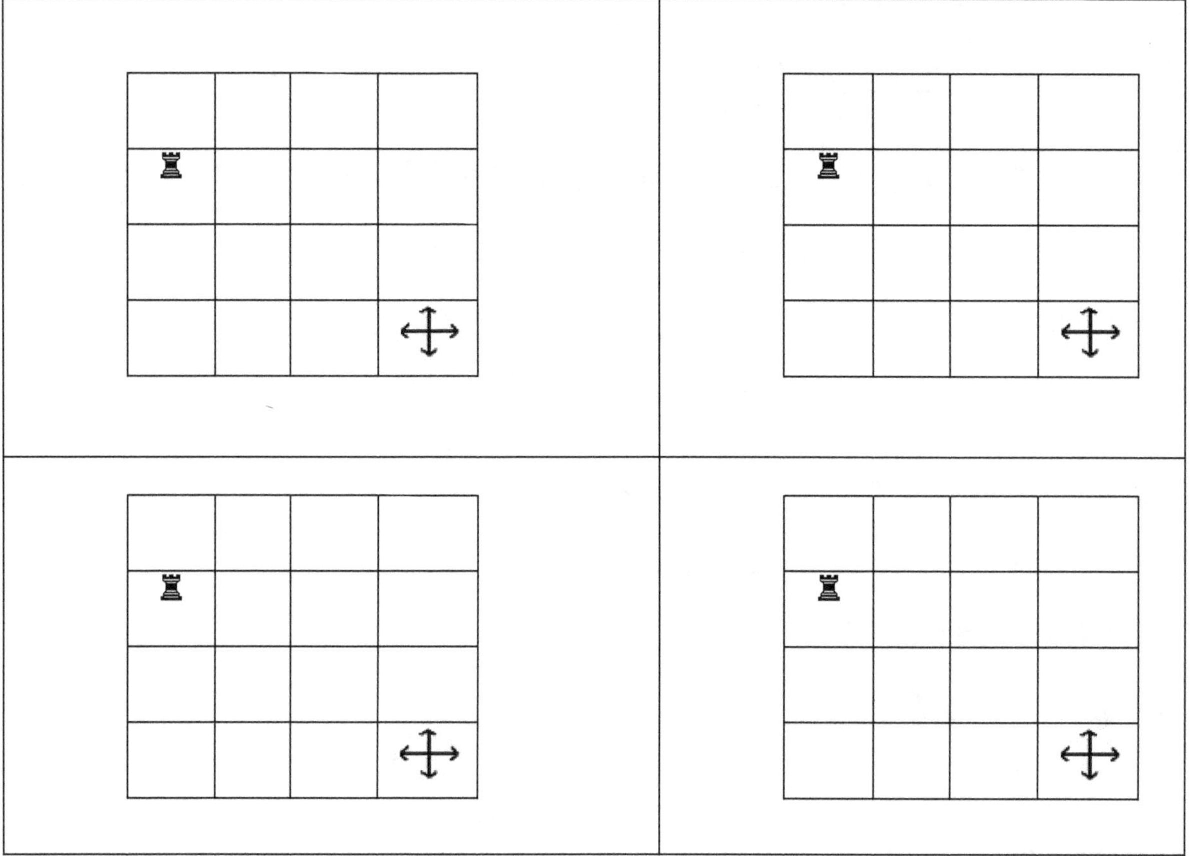

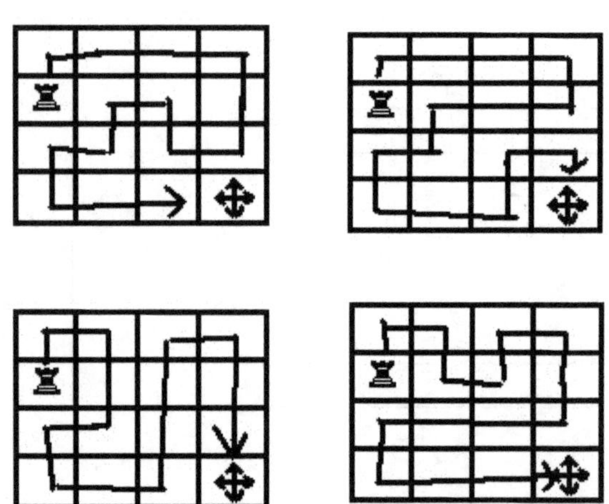

Why buy a Ho Math Chess Franchise?

Intellectual Properties of Ho Math Chess

何数棋谜 - www.homathchess.com

Student's name: ________________ Assignment date: ____________________

Spatial relation and directions

	Direction	Direction	Direction	Direction
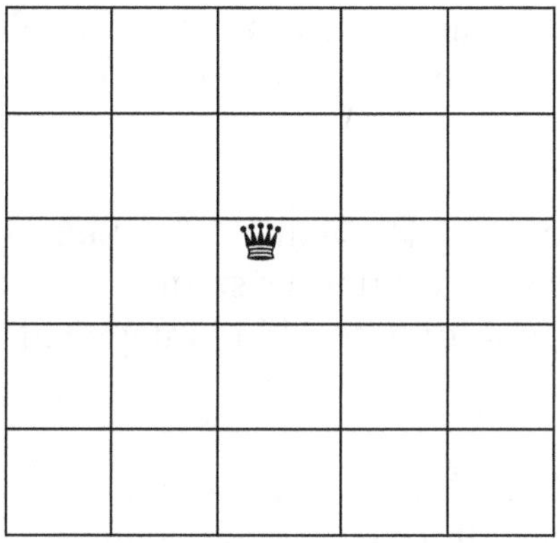 ♛	 Colour red 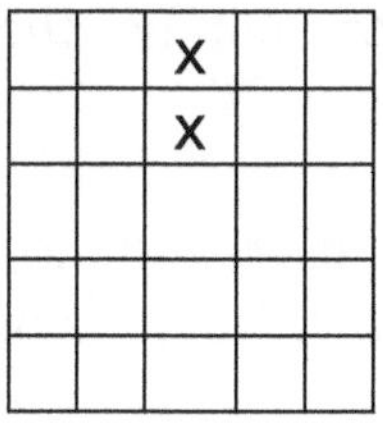 x x	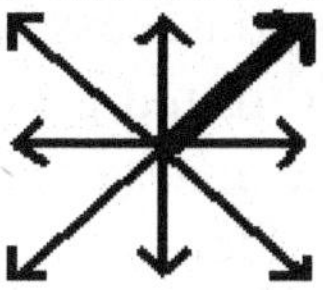 Colour green x x 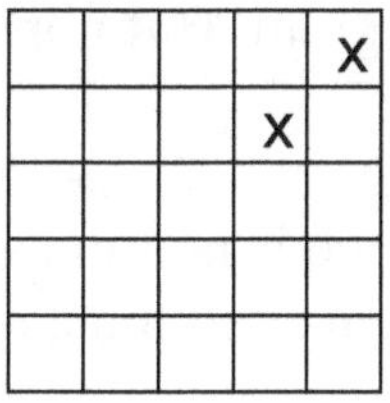	 Colour blue x x 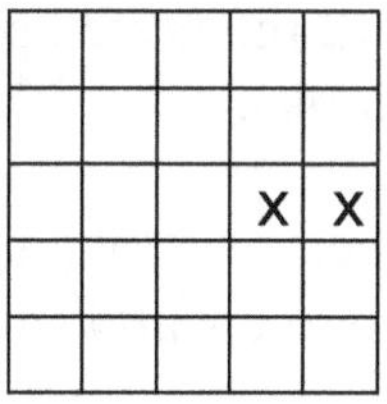	 Colour black x x 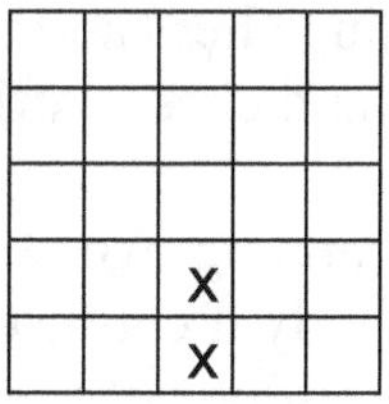
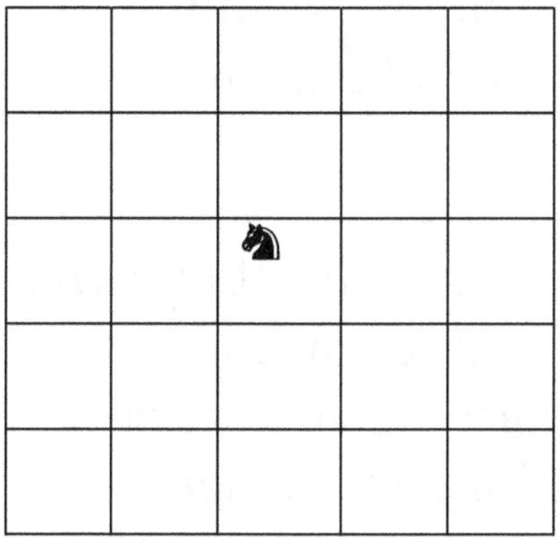 ♞	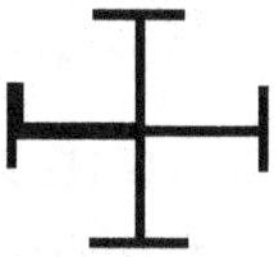 Colour red x 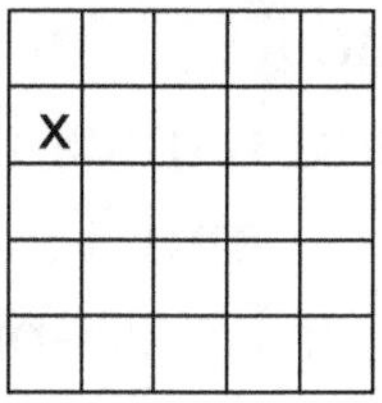	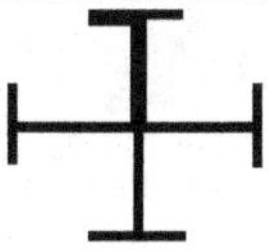 Colour green x 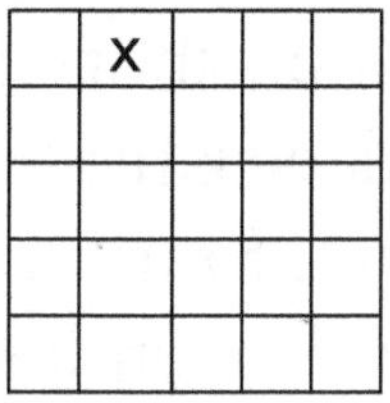	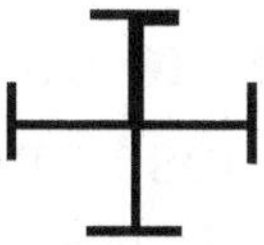 Colour blue x 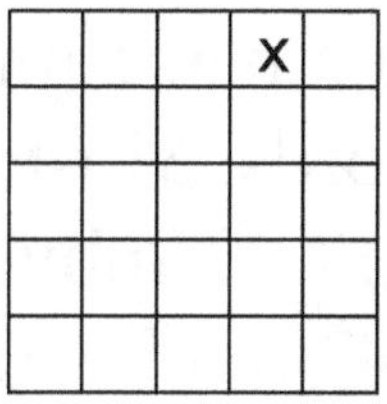	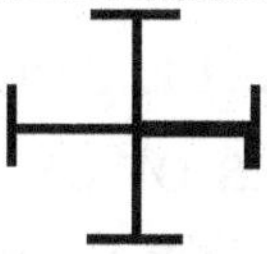 Colour black x 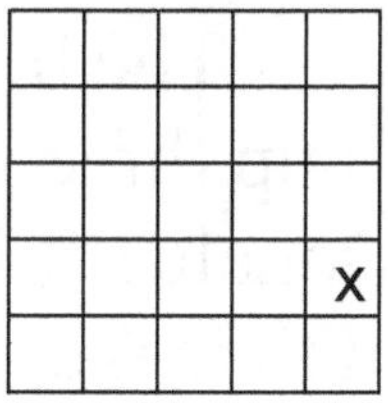

Student's name: ________________ Assignment date: ___________________

The Importance of Using Ho Math Chess Brand Name Only

You mustn't take the first wrong step when creating a business or a business name. As human nature, lots of people would think to use their existing business name or their name and just get Ho Math Chess materials or add Ho Math Chess as an additional product to their brand name, at the risk of being thought as self-promoting myself, this is a mistake made in starting up a business. I have seen cases like this before, so I am blunt but honest about the fact.

Ho Math Chess is getting popular, and if you do not take the stand of promoting Ho Math Chess, then your local competitors will. In Vancouver, we have many places now teach chess after we started teaching chess and math integrated program. But none of them has the capability of teaching integrated math and chess workbooks.

Co-branding two business names into one is not an easy task, and most of the time, it ends up in failure, especially it confuses parents and students and most of the time, the owner will have a prejudice in favour of one of them.

Is it such a good idea to co-brand Ho Math Chess with another existing business or even choose another new business name instead of Ho Math Chess? **The answer is no.** The reasons are as follows:

(1) The brand name must be short, and Ho Math Chess by itself already takes up three words. Most successful brand names have terse names such as McDonald's®, KFC®, Subway®, Starbucks®, and Tim Horton® etc. It is not good to make "Ho Math Chess" even longer by co-branding with other businesses.

(2) We answer "Ho Math Chess" when people phone us, so it is not good to make this greeting and introduction message even longer. It confuses parents and gives the impression that the owner is not very professional since he runs many businesses and wears too many hats. Image ABC learning centre will answer "ABC and Ho Math Chess", what is this?

Student's name: _______________ Assignment date: __________________

(3) Ho Math Chess is a registered trademark name, which stands out by naming "math chess". The name Ho Math Chess itself is like a piece of magnet, which attracts children to call their parent's attention. Children and parents will get confused if Ho Math Chess is mixed with another name. It also clouds the meaning of "math and chess" when mixed with another business name. The value and message of "math and chess leader" are not being brought to parents' attention when Ho Math Chess is mixed with another name. We have kids telling their parents to come to Ho Math Chess just because we have the store name "math chess".

(4) Ho Math Chess is being promoted as an international child education franchise, and its trademark will increase in value when we get more franchisees to join us worldwide. It will become a loss if you have not been using Ho Math Chess as your business name.

Student's name: ________________ Assignment date: ________________

(5) Ho Math Chess, the name itself will become an asset in the future for your business if you have been promoting it. It also builds the right image and brings business when you mention that Ho Math Chess is an international child education franchise with worldwide locations over 20 years of tutoring history in Canada.

(6) When Ho Math Chess is co-branded with another name, not only it might cause to loss of the use of the trademark in your city, it also clouds the message, which "Ho Math Chess" is trying to convey to clients.

(7) Ho Math Chess invented and created an innovative chess teaching set called Ho Math Chess Teaching Set, and this set is circulating worldwide, carrying the Ho Math Chess trademark. How much one has to pay in a marketing budget to get this kind of worldwide exposure?

(8) A publication entitled Frankho ChessDoku has been published by Ho Math Chess and is being circulated the world to promote the brand name.

(8) If the brand name of Ho Math Chess has not been vigorously promoted, then how are you going to convince others to join you as a family member of Ho Math Chess?

(9) You do not get the benefits of www.mathandchess.com been ranked as the number one page on Google®, MSN® or Yahoo® since you do not refer parents to www.mathandchess.com, which has video clips, over 50 articles and over 100 testimonies. You lose business as a result of not being able to use the advertising power of www.mathandchess.com.

(10) You do not get marketing advice from Ho Math Chess headquarters since you do not want to show your flyers and brochures to the headquarters to get a second opinion when you do not use the brand name Ho Math Chess.

Student's name: ________________ Assignment date: ___________________

第 10645081 号

商标注册证

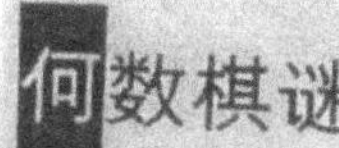

核定服务项目(第 41 类)

学校（教育）；书籍出版；教育考核；教育；在计算机网络上提供在线游戏；培训；广播和电视节目制作；在线电子书籍和杂志的出版；讲课；电子桌面排版（截止）

注 册 人 何数棋谜培训中心有限公司
HO，MATH AND CHESS LEARNING CENTRE INC.

注册地址 加拿大不列颠哥伦比亚省温哥华市滑铁卢街 2586 号
2586 WATERLOO STREET VANCOUVER，BC V6R3H5 CANADA

注册有效期限 自公元 2013 年 05 月 14 日 至 2023 年 05 月 13 日

局长签发 许瑞表

中华人民共和国国家工商行政管理总局商标局

Student's name: ________________ Assignment date: ____________________

Why Ho Math Chess is unique and stands out above the crowd?

(What would destroy Ho Math Chess brand name?)

Many organizations or chess coaches in the world teach math and chess together and under one typical roof. There are also many books published which teach IQ puzzles. Searching on the internet, I found some math, chess, or puzzles teaching-related sites as follows:

1. http://chess-math.org/

The above organization just teaches chess.

2. http://www.philipmathchess.com/en/index.php

The above organization teaches math and chess under one roof but no integrated workbooks.

3. http://www.123mathandchess.com/

The above organization teaches math and chess under one roof but no integrated workbooks.

4. http://www.chesskids.com.au/about/

The above organization is trying to franchise chess worldwide.

5. www.puzzlesfundome.com

The above organization is a daycare centre using puzzles.

Student's name: ________________ Assignment date: ___________________

Browsing through the above sites, you will find what they are doing is basically to teach multiple subjects under one roof, and there are no integrated math and chess workbooks to link these multiple subjects so that students could work on them. So if you now run a learning centre just like one of the above organizations, you can be "easily" replaced by someone who opens a similar learning centre next door. It means that you are setting up a run-of-the-mill type of learning centre, and it has no own culture and no uniqueness.

Ho Math Chess is a culminated product of 20 years of tutoring experience with three different business entities: math, chess, and puzzles, all combined under one roof by means of integrated workbooks using our unique copyrighted intellectual properties. So when you join Ho Math Chess, you are buying three tutoring businesses: Math, Chess, and Puzzles.

The copyrighted intellectual properties of Ho Math Chess owns are listed as follows:

Student's name: ________________ Assignment date: ___________________

The most important identification of Ho Math Chess is our training chess set. These Geometry Chess Symbols (GCS) which are extensively used in our workbooks: Pre-K and Kindergarten Math, 3 by 3 Frankho Puzzles, 4 by 4 Frankho Puzzles, Addition workbook, Subtraction workbook, Junior Magic Chess and Math Puzzles workbook and many of our other mathematical IQ puzzles workbooks.

A serious mistake is being manifested by not following the Ho Math Chess franchise standard to use the Ho Math Chess training set since you will lose Ho Math Chess unique identification to use GCS to link chess, math, and puzzles.

Any learning centre could replace your existing Ho Math Chess by hiring a chess master teacher and using a stand-up traditional chess set if you did not follow Ho Math Chess's franchise standards.

The sympathy would be hard to come by if one day, you found that someone opens a similar learning centre to yours in your area because you did not follow the Ho Math Chess franchise standard.

Our unique math, chess, and puzzles integrated workbooks.

Math is a world universal school curriculum subject and is being taught by every school globally, so when chess and puzzles are added to it, math becomes an interesting subject for children. So why you want to dilute the world-famous brand name of Ho Math Chess by adding other subjects or products (such as other company's products or drawing classes etc.) and de-emphasize its specialized teaching of math?

Student's name: ________________ Assignment date: ____________________

The result is you lose the recognition of the brand name of Ho Math Chess.

Ho Math Chess is selling its brand name, and you must be careful not to dilute Ho Math Chess's strong brand name or weaken Ho Math Chess brand name by just adding on other subjects or products without careful consideration and approval from Ho Math Chess.

Ho Math Chess wants to continue strengthening its brand name that is Ho Math Chess = MATH + CHESS+ PUZZLES, and that is it.

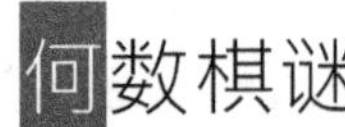

Student's name: ________________ Assignment date: ____________________

How is Ho Math Chess different from others?

There are many different kinds of math learning centres, but by and large, they can be classified into two categories according to their functionalities. Some are created to help students increase their school math marks, and others are created for boosting testing results. Ho Math Chess, founded by Mr. Frank Ho, a Canadian certified math teacher, had a background story to tell how Ho Math Chess got started. It was not initially created for increasing students' school marks, nor was it created for preparing for entrance exams such as SSAT, SAT etc. Ho Math Chess was created because Frank's son was interested in chess at age five, and Frank got involved in teaching his son chess. Ho Math Chess was started because of Frank's interest and research in math and chess integrated subjects. Amanda later joined Frank, and the Ho Math Chess research team has created many world's first math and chess integrated workbooks and a new chess teaching set by using the copyrighted technology.

How Ho Math Chess is different from others can be summarized as follows.

	Ho Math Chess	Others
Has a founding philosophy, culture, and story to tell.	yes	no
Taught own kids used the same teaching method.	yes	no
Has new, innovative, and copyrighted intellectual property.	yes	no
Can increase school math marks	yes	yes
Can prepare for entrance tests	yes	yes
Created teaching materials using copyrighted technology	yes	no
Integrate chess into teaching	yes	no
Integrate puzzles into teaching	yes	no
Learning can boost IQ scores	yes	Not sure
Chess can be played as a lifetime hobby	yes	no
Proven results in the teaching method.	yes	Not sure
Integrate computation, chess, word problem, puzzles all in one worksheet using copyrighted technology	yes	no

Student's name: ________________ Assignment date: ___________________

Enriching Math Using Chess

This article is published by the official Journal of the British Columbia Association of Mathematics Teachers, British Columbia, Canada (Vector, Summer 2006, Volume 47, Issue 2, page 44).

Please click the following web address to view the entire article.

http://www.scribd.com/doc/208007379/Enriching-Math-Using-Chess

Background

I started to teach chess to my son when he was five years old and soon noticed that the relationship between mathematics and chess is one of those generally presumed truisms. I was not able to find a book of junior-level math and chess hybrid problems for my son to work on, however. In 1995, after seriously looking into the possibility of writing one myself, I pioneered the idea of integrating chess symbols and their values into math. For the first time, elementary students were able to visualize how math and chess concepts are related. The idea of using chess symbols and their point values and chess moves directly in puzzles characterizes what I believe is a breakthrough that permits a true integration of math and chess since the puzzles are no longer limited to only traditional chess puzzles.

Later, I created worksheets with a 2-column system; chess on one side and math puzzles. Students are given the opportunity to see the relationship between math and chess side by side. I also used chess symbols directly in arithmetic operations. The effect is that simple, one-step questions instantly become abstract and symbolic multi-step questions requiring children to analyse the problem and take necessary steps to understand the abstract concept before coming up with a solution – excellent training for critical thinking problem-solving.

I have been fortunate to have the opportunities to personally field-test these problems at my learning centre since I have been teaching math from kindergarten to grade 12 over the past 20 years. This unique experience has allowed me to obtain a wide spectrum of feedback from different student backgrounds. This article will attempt to show how different kinds of mathematical chess puzzles can be produced and their potential benefits for learning outcomes.

Chess benefits children

Why chess fascinates children? Dr. Montessori observed that younger children were intensely attracted to sensory development apparatus. Chess, being hands-on and multi-sensory, involves coordination between eyes, brain and hands in multi-direction and embodies concepts that are non-linear when compared to most video and computer games.

Student's name: ________________ Assignment date: ____________________

As young as five, my son was thrilled to capture my chess pieces as his reward, that perhaps was the sensory experience that kept him hungry for more games.

In addition to having fun, playing chess has been shown to improve cognitive and critical thinking skills, reasoning and problem-solving abilities, focusing, visualizing, analytical and planning skills. These conclusions have been backed up by educational research papers (1).

How are my puzzles different from others?

In the past, many chess puzzles have been published, and as for strictly chess problems, the standard has long been set by Sam Loyd, the "Puzzle King", composer of some of the most paradoxical, almost phenomenal, chess problems. The most recent mathematics and a chess-related book entitled Mathematics and Chess has 110 entertaining problems and solutions. (2) Almost all of these published puzzles, old and new, are related to the moves of chess pieces and the majority of them are considered too difficult for most elementary students to comprehend.

A math textbook series called "Challenging Mathematics" (3) has chess as part of a logic section, but the chess content itself is stand alone and is not integrated with any math concepts or math problems.

I used chess symbols, chess values, chess moves, chessboard, algebraic notation, chess set-up, attacking and defending counting, the order of exchanges, and chess rule etc., to create mathematical chess puzzles.

The fundamental difference between my mathematical chess puzzles and those traditionally published is that chess symbols, moves, and values are integrated into math to create problems in patterns, logic, geometry, counting paths, relation, arrangement, numeration, and even data management. In this way, children from pre-kindergarten through elementary school, while learning chess, are provided with many opportunities to explore mathematical puzzles by making use of the very basics of chess knowledge.

The puzzles are designed to enhance math ability using chess as a teaching tool. It is not intended to substitute for instruction in school math but rather to serve as enrichment or supplemental material.

Children learn best by playing games, Math + Chess = a fun way of learning math.

Student's name: ________________ Assignment date: ___________________

How chess and math are integrated

The creation of math and chess problems require one to have thorough understanding of chess knowledge and also the school mathematics curriculum at each grade level, Only one processes these qualifications and creative mind, can a meaningful mathematical chess puzzle be created. Integrated chess and math problems are created using the following principles:

(1) Chessboard and chess pieces.

The chessboard is symmetric in main diagonals in terms of its colour. The chessboard is made of four identical small boards if it is divided by one horizontal line and one vertical line going through the centre. The set-up position of chess pieces is symmetric between Black and White. The chess pieces' set-up position on either side is a palindrome, excluding the king and queen.

The ranks and files are related to coordinates. When a piece is being attacked or defended, it requires some arithmetic calculations in terms of the number of attacking and defending pieces, and this is the first lesson a child would learn in counting numbers.

(2) Chess moves

Rook's move is a slide motion (left/right, up/down) in geometry. The between moves of rook before reaching the destination is similar to the concept of commutative property. For example, to reach from a1 to h1(7) = a1 to c1 (2) + d1 to h1 (5) = h1 to d1 (5) + c1 to a1 (2).

Student's name: ________________ Assignment date: ____________________

To figure the "best" move, one needs to find out all possible paths. The similar concept in math would be to use the factor tree to find out the prime factors of a number, for example, to find out the prime product of a number 32. One could use a factor tree to do it.

How to checkmate an opponent? If the rook is at $a1$ and is free to make moves along with file a and rank 1, what has to be considered before moving? To see if there any opponent's pieces intersecting with rook would be like to find out what would be y when $x = 1$? The checkmate positions are actually the intersections of ranks or files, which are very similar to the concept of solving equations by using the graphics method.

To find the common squares (squares both pieces could control) is similar to the idea of finding common factors of two numbers.

One would think that chess perhaps has nothing to do with fractions since all moves are all in whole numbers? Why queen is the most powerful piece and usually we move pieces toward the centre? They all have something to do with the ratio $a/64$ where a is the number of squares under a queen's control.

When chess players check possible moves, the view encompasses a circular locus through 360^0 movement. For example, when checking a rook's possible moves, a player scans the following angles: $0^0, 90^0, 180^0, 270^0$. In other words, the rook's move is equivalent to rotate the rook in four directions. The same rotation concept is true for bishop, queen, and king.

(3) Chess Symbols and Chess Values

The Use of Chess Symbols as Constants

English letters such as x, y, z, ... are normally used to represent unknown numeric values. These unknown letters are also called variables, and they normally do not have singly defined values. On the other hand, each chess symbol has definitely defined meaningful point value that is related to each piece's strength in the chess game. The point system is the static value of a piece and generally serves as a guide to making trades with the opponent. Take a look at the following example.

Student's name: ________________ Assignment date: ___________________

Let $x = 1$, $y = 3$ then $x + y = 1 + 3 = 4$

In the above particular example, x is 1 and y is 3 but x does not always have to be 1, nor does y always have to be 2.

If we use chess symbols in the above example, we get

♟ + ♝ = 4

The above pawn and bishop have specifically defined values 1 and 3 respectively and will not change their values just because the problem is different. In other words, they are constants.

By contrast, in algebra, students would substitute x or y with different values when given values are changed. In other words, in algebraic equations, the values of x and y change in terms of the particular problem. To compare the substitution values for chess symbols and algebraic variables, we realize that there is a difference in the chess symbols substitution - it is intuitive for children since the value of each chess symbol is pre-defined and hence has an implicit meaning to them.

The idea of using chess symbols is to teach children how to transfer from a concrete object into an abstract concept. For example, a concrete object (like a bishop chess piece) could be associated with a symbol ♝, which in turn could be substituted with a value of 3. This process of learning and thinking skill learned is in line with the concept of Montessori teaching.

To use animal figures or any other symbols such as x, y, z in mathematical chess puzzles would be less meaningful to children when compared to the use of chess symbols in math and chess integrated puzzles problems. Children do not get confused on the use of chess symbols, nor will they become handicapped when they learn the variables in high schools just because they learn substitution at a young age. The chess symbols are only used as "pictograms", i.e. as representations of values. They understand the spaces in an arithmetic expression where these chess symbols occupy should have been numbers in the first place when used in arithmetic operations.

Student's name: ________________ Assignment date: ____________________

The other reason for using chess symbols in mathematical chess puzzles is the chess symbols themselves representing movements, and coincidently some of the directions of movements resemble some arithmetic operators. For example, the rook can move up and down and left-right, and thus its trace of capable moves looks like a + sign.

Each chess symbol has a specially defined direction of the move and these directions of moves are "embedded" with each piece. I have taken advantage of chess pieces moves and defined them as follows:

Addition/Subtraction = Rook (Could also be queen or king)
Multiplication = Bishop (Could also be queen or king)
Division = King (Opposition of 2 kings)

Chess Values Used

The values of chess symbols are the same as the ones used in the Teaching Manual published by the Canada Federation of Chess (4).

The cancellation technique to counting the points of chess pieces would be very similar to the concept of subtraction property of an equation.

Chess symbols and values are integrated into arithmetic operations to create a new type of problem. My purpose in using chess symbols is to create more interesting questions and encourage children to think ahead and use spatial-temporal reasoning to solve problems.

Each chess piece has been assigned a different point (value). For example, the following are values assigned to chess symbols.

♔ (king) = 0 point	♘ (knight) = 3 points	♖ (rook) = 5 points
♙ (pawn) = 1 point	♗ (bishop) = 3 points	♕ (queen) = 9 points

My experience of using chess values in teaching arithmetic operations is very positive. Elementary students who have not learned variables but have worked on my worksheets using chess symbols have absorbed the concept of algebraic variables or substitution in a natural and intuitive way.

Student's name: ________________ Assignment date: ___________________

There is no need to explain the concept of variable other than mention the values of chess pieces, for example,

♜ + 5 = _____

Examples

Listed below are a few examples I created to show the relations between math and chess.

Example 1 Addition and Subtraction

□ ――― + ♝ ♛ □ − ♜ ――― ♛	□ ――― + ♜ 6 □ − ♜ ――― 6

The above problem is the one I designed to go against conventionally designed worksheets, which is always from left to right in a linear fashion. One could work out the above problem from the bottom to top and then from top to down in multi-direction.

Student's name: ________________ Assignment date: ____________________

Example 2 Multiplication

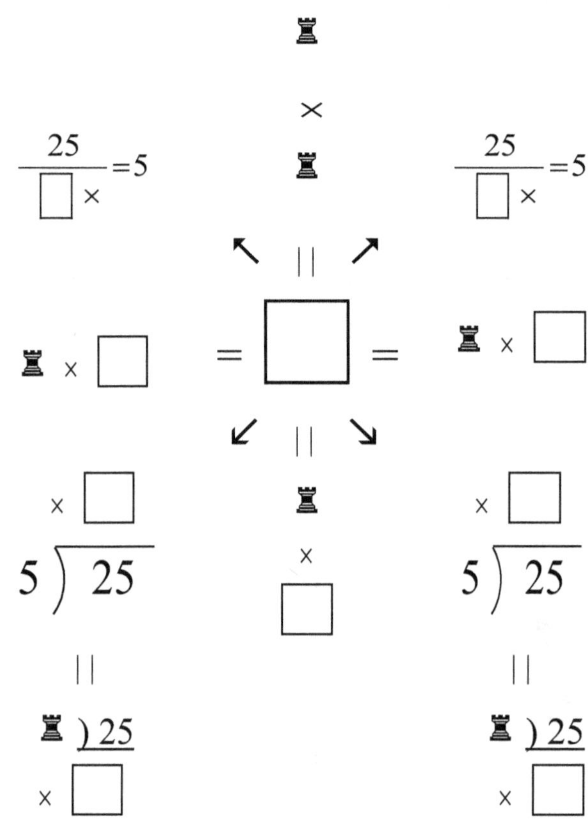

The above problem is created with the mind that children do not really learn math in a sequential way of addition, subtraction, multiplication, or division in real life. So here, you could see how I incorporate the idea of multiplication in different formats of computing. The purpose of this worksheet is to learn multiplication but expose children to think about how multiplication could be written in different ways. A simple multiplication problem is changed to a two-step problem in multi-direction, multi-operation, and multi-concept learning.

Student's name: ________________ Assignment date: ___________________

Example 3 Logic

Chess Symbol	Logic Training
A new Chess Symbol is defined as follows: Chess figurines / Chess symbols: ♔ — ÷ (Opposition) ♖ — ┼ ♘ — L ♗ — X ♕ — ✳ ♙ — ⩛	In the following equation, observe the chess symbols on the left and fill in each ◯ with a number. If ┼ + ┼ = 10 then ÷ + ┼ = ◯ ┼ ✳ + ┼ ÷ ◯ ÷ ✳

Use the above Chess Symbol table; find the following pattern:

Z, ÷, O, ⩛, T, L, T, X, F, ┼, ______, ✳

Use the above Chess Symbol table; find the following pattern:

0, ÷, 1, ⩛, ___, L, 3, X, 5, ┼, ___, ✳

The above problem demonstrates that not only children will not get confused on traditional chess symbols used in an arithmetic expression in my workbook; they could be led to use additional "creative" chess symbols and are able to solve them correctly. This problem is suitable for grade 3 and above students.

Most students could not solve the last 2 puzzles, but they appreciate the sophistication of the problems when I explained to them the logical relationships. The last problem is the one using a combination of chess symbols and their values.

Student's name: ________________ Assignment date: ____________________

The use of chess values is much like the use of monetary values. When chess or money figures are seen by children, they both represent some pre-defined meaningful values. The following is an example where the values of chess pieces could be monetary values and "Total Points" could be the sum of the total money.

Example 4 Table Values

Fill in the different number of chess pieces to come up with each total.

Number of ♟	Number of ♞	Number of ♜	Total points
1	1	1	9
3	2	0	9
0	3	0	9
☐	☐	☐	10
☐	☐	☐	10
☐	☐	☐	11
☐	☐	☐	12
☐	☐	☐	13
☐	☐	☐	14
☐	☐	☐	15

Student's name: ________________ Assignment date: ____________________

Example 5 Equation

The following examples demonstrate how chess symbols and chess values are integrated with arithmetic operations to facilitate critical thinking skill.

♛ + ♞ + x = 54
x = _____

Student's name: ________________ Assignment date: ____________________

Example 6 Addition and Subtraction, If Then - Else

	10	−	♚	=	□	
+	5	+	♚	=	□	+
	□				□	

If 10 + ♜ = □, then 9 + ♜ must be □.

If ♜ + 10 = □, then ♜ + 9 must be □.

Student's name: ________________ Assignment date: ___________________

Example 7 Cross Multiplication

♟ = □   8 = □

×   ↘ ↙ × ↙ ↘   ×

□   □

□ + □ = 6

Why buy a Ho Math Chess Franchise?

Intellectual Properties of Ho Math Chess

何数棋谜 - www.homathchess.com

Student's name: ________________ Assignment date: ____________________

Example 8 Multiplication and division

♛ × 2 ——— 18 ÷ 2 = □	♛ × ♜ ——— □ ÷ 5 = □
♛ × ♜ ——— □ ÷ 9 = □	♛ × ♛ ——— □ ÷ ♛ = □
♛ × ♞ ——— □ ÷ ♞ = □	♛ × ♞ ——— □ ÷ ♛ = □
♜ × ♞ ——— □ ÷ ♞ = □	♜ × 8 ——— □ ÷ ♜ = □

Will the above operations cause confusion to children, the answer is no. ♛×♞ will not make sense if it is explained literally as a queen times a knight. However, if it is translated into numerals 9 times 3 then the product must be 27, which is very logical, and children understand that they are working on 9 × 5, not the product of ♛×♞.

A similar type of logic question is as follows: If 2 # 3 is defined as 2 + 2 × 3, then what is 3 # 4? Normally 2 #3 will not make any sense since it is not a valid arithmetic operator, but if we define it clearly, then it becomes workable.

Student's name: ________________ Assignment date: ___________________

Example 9 The following is a puzzle that requires the knowledge of chess moves.

Filling in □ with a chess piece	Geometric shapes
♙	
♖	
□	
♗	

Student's name: ________________ Assignment date: ____________________

Example 10 Use chess moves to solve the following puzzle.

	?		14	
?				21
		♘		
?				28
	42		?	

On the first look, lots of students are not able to solve it; why? Students are so used to do computation from left to right, and this question has to be solved in an unconventional direction.

Example 11 Use chess symbol moves to solve the following puzzle.

If 2 ♜ 3 = 5 then 2 ♝ 3 is = ____

Surprisingly, some of my students have no trouble to solve the above puzzle.

Student's name: ________________ Assignment date: ____________________

Example 12, Use chess symbol moves to solve the following puzzle.

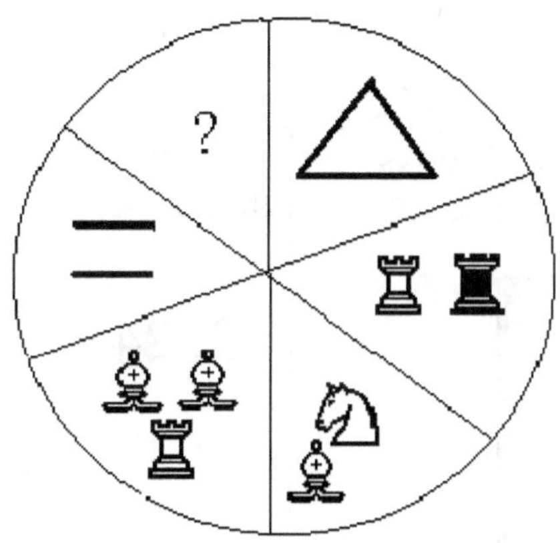

Example 13

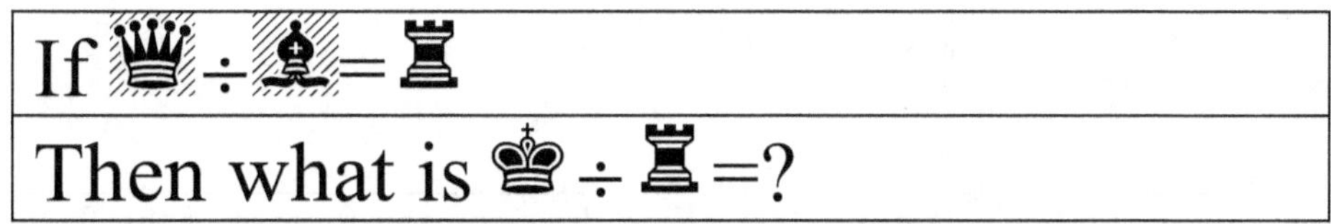

This problem cannot be solved using chess values; students tried it and knew it. So what is the trick behind the idea of this puzzle?

Student's name: ________________ Assignment date: ____________________

Example 14

Rook Path

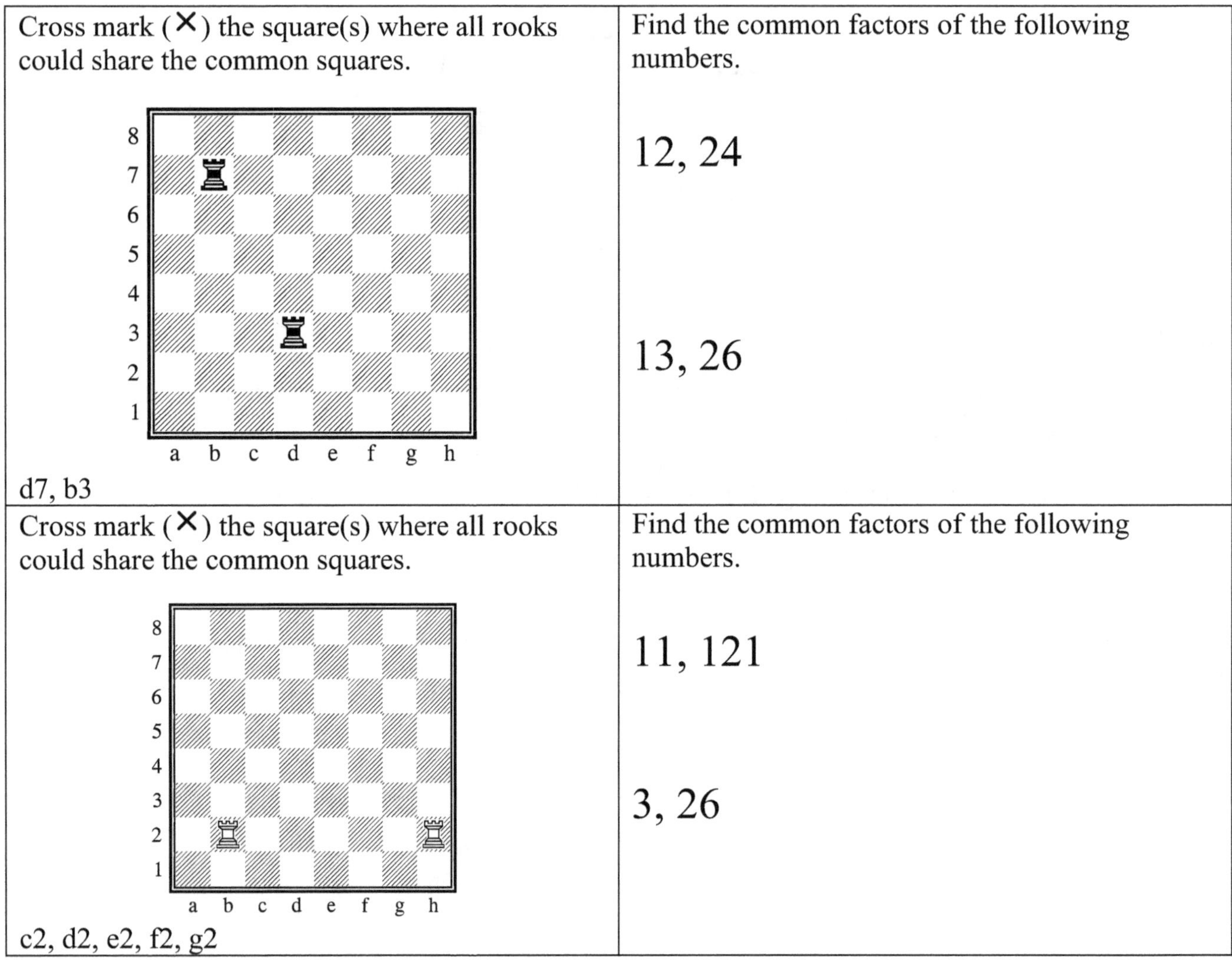

Cross mark (✕) the square(s) where all rooks could share the common squares. d7, b3	Find the common factors of the following numbers. 12, 24 13, 26
Cross mark (✕) the square(s) where all rooks could share the common squares. c2, d2, e2, f2, g2	Find the common factors of the following numbers. 11, 121 3, 26

Chess moves problems could be applied to math concepts. I give the above example to demonstrate the idea with chess problem on one side, and the other side is the math problem.

Student's name: ________________ Assignment date: ___________________

Summary

I have found the idea of using chess symbols very helpful for elementary students – they could learn to solve mathematical chess puzzles using their chess knowledge.

Many children could not play competitive chess well when against someone but feel very proud that they could solve mathematical chess puzzles. Mathematical chess puzzles provide some children with additional opportunities that they could challenge themselves. For this reason, I give prizes to chess winners and also to puzzles solvers.

The most interesting in using chess symbols is that the chess symbols themselves not only possess pre-defined values but also have the implied meaning of movements and these two special characters allow me to create some very interesting mathematical puzzles with pizzazz.

By using chess symbols, a simple one-step arithmetic problem could become a multi-step problem. As a result, chess symbols and values offer children more opportunities to work on another type of question, which could stimulate children's brain cells and improve their problem-solving ability. So the benefits of working on these types of problems are double-edged- improving chess knowledge and also mathematical problem-solving ability.

The mathematical chess puzzles created by me are not just mechanically substituted numbers with chess symbols. Many mathematical chess puzzles created also involve pattern, sequence, geometry, set theory, and logic etc. In other words, the integration is much diversified and also involves multi-direction visualization. I would give the following examples to demonstrate how chess symbols and values are presented in the following pattern-like puzzles with the multisensory approach using multi-direction.

Student's name: ________________ Assignment date: ___________________

Example 15 Find values to replace? or fill in □.

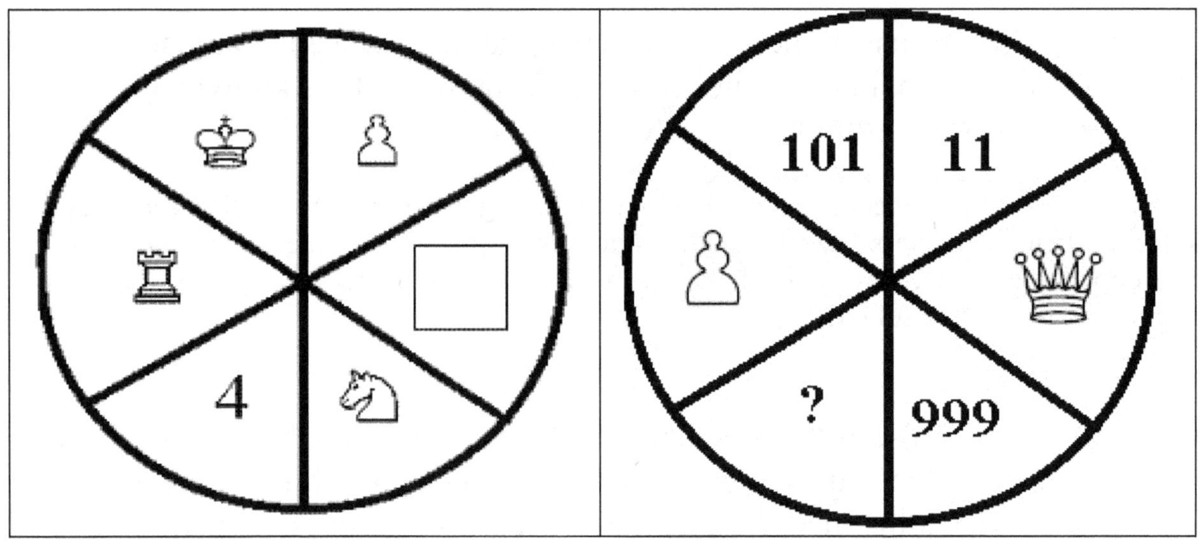

Reference

(1) Chess in Education Research Summary, Dr. Robert Ferguson. Details see below http://www.quadcitychess.com/benefits_of_chess.html
(2) Mathematics and Chess, by Miodrag Petkovic, Dover Publications, Inc. Mineola, New York, 1997
(3) Challenging Mathematics, Cheneliere/McGraw-Hill, Michael Lyons, Robert Lyons, 1995, Montreal, Quebec, Canada
(4) Chess Teaching Manual, IM Tom O'Donnell, the Chess Federation of Canada.

Student's name: _______________ Assignment date: ___________________

news

Tutor makes learning more fun

Chess moves provide mathematical insight

By Mark Hasiuk
Staff writer

"If kids play chess at an early stage, six or seven years old, they in fact are learning mathematics principles they won't learn until Grades 4 and 5."
—Frank Ho

A VANCOUVER MATH TUTOR is using the ancient game of chess to help students checkmate problems with math.

Frank Ho is the founder of Math + Chess, a tutoring company headquartered in Kerrisdale that offers once-a-week sessions to students from Grades K-12, although he mostly deals with elementary school children.

Ho said his unique teaching method combines the elements of fun and competition with the mathematical principles required to play chess.

"The goal is to checkmate your opponent, but to do that you need to devise patterns and visualize a chess diagram and come to a numerical conclusion," said Ho, adding that students compete against each other or an instructor.

Ho said young chess players develop math skills more advanced than what they learn during their regular school studies.

"If kids play chess at an early stage, six or seven years old, they in fact are learning mathematics principles they won't learn until Grades 4 and 5," he said.

"In school, children learn in simple one line patterns like 2, 4, 6, 8, from left to right or sometimes vertically. When they get older they get into fractions and cross multiplying, but we train them to think in a multi-directional way at an early age."

Ho emigrated from Taiwan in 1978 and earned a masters degree in statistics before becoming a statistical consultant at UBC. He started Math + Chess in 1995, after watching his son—who is studying medicine at the University of Manitoba—become Canada's youngest chess master at the age of 12.

Ho studied the relationship between math and chess in textbooks and several academic journals, and became convinced that combining the two disciplines could benefit young students.

"Math problem-solving can get boring because you are basically just competing with your brain," he said. "This is a hands-on approach of friendly competition where students can see a result before their very eyes."

Ho's Kerrisdale learnin centre now tutors approx imately 200 kids a mont at $34 for a two-hour ses sion.

The success of hi Vancouver operation al lowed Ho to franchise and there are now Mat + Chess outlets in Burn aby, Richmond, a hand ful in the United State and one each in India an Mexico.

Mary Zahrai enrolle her six-year-old daughte in Math + Chess so sh would have an easier tim with her regular studies.

"Her problem-solv ing has really improved, said Zahrai. "I thought i might be too complicate for her, but it's been reall good for her brain. She i doing very well with he school math, and I kno she likes the challenge o competition when play ing chess with the othe kids."

Student's name: ________________ Assignment date: ____________________

Enriqueciendo las Matemáticas con el Ajedrez

Frank Ho

Profesor del Centro de Aprendizaje Ho -Math and Chess™ de Vancouver

Perfil

Empecé a enseñar ajedrez a mi hijo cuando tenía cinco años de edad y pude observar que la relación entre las matemáticas y el ajedrez es una de aquellas grandes y obvias verdades. No logré hallar un libro de problemas para niños que combinara matemáticas y ajedrez y que pudiera ser útil para mi niño pequeño. En 1995, luego de reflexionar seriamente sobre la posibilidad de yo mismo escribir aquel libro, fui pionero en la idea de integrar a los símbolos de ajedrez y sus respectivos valores, con las matemáticas. Por primera vez, los estudiantes de educación primaria, serían capaces de visualizar la forma en que se relacionan los conceptos de matemáticas y ajedrez. La idea de utilizar directamente los símbolos de ajedrez y sus valores con puntos, así como las movidas del ajedrez en rompecabezas, pone de manifiesto mi visión sobre lo que considero es una innovación que permite una real integración entre matemáticas y ajedrez, dado que los rompecabezas ya no se limitan a aquellos tradicionales rompecabezas de ajedrez.

Luego, diseñé un sistema de hojas de trabajo que contienen dos columnas: a un lado ajedrez y al otro rompecabezas de matemáticas. Se le da al estudiante la oportunidad de ver la relación entre las matemáticas y el ajedrez, una al lado del otro. También hice uso de símbolos en las operaciones aritméticas. El efecto es así de simple: preguntas de un paso se vuelven, al instante, preguntas de múltiples pasos, abstractas y simbólicas, que llevan a los niños a analizar el problema y tomar los pasos necesarios para entender el concepto abstracto antes de dar con la respuesta- un excelente entrenamiento para el desarrollo del pensamiento crítico y la solución de problemas.

He sido afortunado al contar con la oportunidad de probar estos problemas en mi centro de aprendizaje, ya que he venido enseñando matemáticas a niños de toda edad, durante los últimos 10 años. Esta experiencia única me ha permitido obtener un amplio espectro de retroalimentación de diversos perfiles de estudiante. Este documento pretende mostrar la forma en que se puede crear una variedad de rompecabezas de matemáticas y ajedrez y sus ventajas potenciales para el desempeño académico de los niños.

Student's name: ________________ Assignment date: ___________________

El ajedrez beneficia a los niños

¿Por que el ajedrez fascina a los niños? El Dr. Montessori observó que los niños pequeños se ven atraídos por instrumentos de desarrollo sensorial. Siendo el ajedrez un instrumento práctico y multidireccional, logra la coordinación ojo-cerebro-manos y personifica conceptos que no son lineales, si los comparamos con la mayoría de los juegos computarizados y de video. Tan es así que desde los 5 años, mi hijo se emocionaba al ser capaz de capturar mis piezas de ajedrez, mientras que su recompensa, que quizás venía dada por la experiencia sensorial, lo mantenía con sed de más juegos.

Además de lo divertido que es, investigaciones en educación (1) demuestran que el juego de ajedrez estimula el desarrollo de destrezas cognitivas, el pensamiento crítico, habilidades de razonamiento y de solución de problemas, focalización, visualización, destrezas analíticas y de planificación.

En qué se distinguen mis rompecabezas de los demás

Anteriormente, se publicaron varios rompecabezas de ajedrez y en lo que respecta a problemas de ajedrez, Sam Loyd, el "Rey del Rompecabezas", estableció el estándar. El libro más reciente, titulado Math and Chess, tiene 110 problemas y soluciones (2). Casi todos los rompecabezas publicados hasta la fecha, están relacionados con movidas de las piezas y la mayoría son muy difíciles para los niños de primaria.

El libro "Challenging Mathematics" (3) tiene contenidos de ajedrez, como parte de una sección de lógica pero no está integrado con ningún concepto o problema matemático.

Utilizo símbolos de ajedrez, valores de ajedrez, movidas de ajedrez, tablero de ajedrez, anotación algebraica, estructura del ajedrez, contabilización de ataque y defensa, orden de intercambios, regla de ajedrez, etc., para crear rompecabezas de matemáticas y ajedrez. La principal diferencia entre mis rompecabezas de matemáticas y ajedrez y aquellos publicados tradicionalmente es que los símbolos de ajedrez, las movidas, y los valores están integrados a las matemáticas para crear problemas sobre patrones, lógica, geometría, contabilización de rutas, relación, agrupación, numeración, e inclusive administración de datos. De esta forma, a la par que los niños de preescolar y primaria juegan ajedrez, tienen la oportunidad de explorar rompecabezas de matemáticas, solo requiriendo un conocimiento básico de ajedrez.

Student's name: ________________ Assignment date: ____________________

Los rompecabezas están diseñados para mejorar las habilidades matemáticas, haciendo uso del ajedrez como una herramienta de enseñanza. No se pretende con ellos el reemplazo de la enseñanza de matemáticas de la escuela, pero si que sirva como un material complementario y enriquecedor. Los niños aprenden mejor jugando, como en la fórmula: Matemáticas + Ajedrez = una forma divertida de aprender ajedrez.

Cómo están integradas las matemáticas con el ajedrez

Para la creación de problemas de matemáticas y ajedrez se requiere conocer a fondo el ajedrez y también saber el currículo escolar de matemáticas de cada nivel, y solo con estas calificaciones y con una mente creativa podrá crearse un rompecabezas de ajedrez con sentido. Los problemas integrados de ajedrez y matemáticas son creados empleando los siguientes principios:

(4) Tablero de ajedrez y piezas de ajedrez.

El tablero de ajedrez es simétrico en sus diagonales centrales, en relación a su color. El tablero de ajedrez es hecho de cuatro pequeños tableros idénticos, si lo dividimos con una línea horizontal y una línea vertical partiendo del centro. La posición establecida de las piezas de ajedrez es simétrica entre Negro y Blanco. La posición establecida de las piezas de ajedrez en cualquiera de sus lados, es palíndromo excepto el rey y la reina.

Los rangos y filas están relacionadas a coordenadas. Cuando una pieza es atacada o defendida, requiere algunos cálculos aritméticos en términos del número de piezas de ataque o defensa y esta es la primera lección que un niño aprenderá al contar números.

(5) Movidas de ajedrez

La movida de la torre consiste en un movimiento (izquierda/derecha, arriba/abajo) de geometría. Las movidas de la torre antes de alcanzar su destino son similares al concepto de la propiedad conmutativa. Por ejemplo, para llegar desde a1 a h1(7) = a1 a c1 (2) + d1 a h1 (5) = h1 a d1 (5) + c1 a a1 (2).

Student's name: ________________ Assignment date: ___________________

Para lograr la mejor movida, se necesita encontrar todos los posibles caminos, y el concepto similar en matemáticas sería usar el árbol de factores para encontrar los factores primos de un número. Por ejemplo, para encontrar el producto primo de un número 32, uno podría encontrar el árbol de factores al hacerlo.

¿Cómo hacer jaque mate al oponente? Si la torre está en $a1$ y es libre de hacer las movidas a lo largo de la fila a y el rango 1, qué es lo que se debe considerar antes de mover la pieza? El hecho de ver si hay otras piezas oponentes que se intersecan con la torre, será como querer hallar y cuando $x = 1$. Las posiciones de jaque mate son en realidad las intersecciones de rangos o filas, las cuales son muy similares al concepto de resolver ecuaciones usando el método gráfico.

El hallar los cuadrados comunes (cuadrados que ambas piezas puedan controlar) es similar a la idea de hallar factores comunes de dos números.

Uno podría pensar que el ajedrez no tiene nada que ver con fracciones, ya que todas las movidas son hechas con números enteros? ¿Por qué la reina es la pieza más poderosa y usualmente movemos las piezas hacia el centro? Todos tienen que ver con la ratio $a/64$ donde a es el número de cuadrados bajo control.

Cuando el jugador de ajedrez revisa las posibles movidas, la vista que realiza es en forma de un locus circular de 360^0. Por ejemplo, cuando se ven las posibles movidas de la torre, un jugador escanea los siguientes ángulos: $0^0, 90^0, 180^0, 270^0$. En otras palabras, la movida de la torre equivale a rotarla en cuatro direcciones. El mismo concepto de rotación, se aplica para el alfil, la reina y el rey.

(6) Símbolos de ajedrez y Valores de ajedrez

El uso de Símbolos de ajedrez en tanto constantes

Student's name: ________________ Assignment date: ____________________

Letras tales como x, y, z, … se usan normalmente para representar valores numéricos desconocidos. A estas letras desconocidas también se les denomina variables y normalmente no tienen definidos los valores. Por la otra parte, cada símbolo de ajedrez tiene un valor de puntos significativo y definido el cual está relacionado con el peso que tenga cada pieza en el juego de ajedrez El sistema de puntos es el valor estático de una pieza y generalmente sirve como una guía para hacer intercambios con el oponente. Observe el siguiente ejemplo.

Si $x = 1$, $y = 3$, entonces $x + y = 1 + 3 = 4$

En el ejemplo anterior, x es 1 y y es 3 pero x no siempre tiene que ser 1 ni y tiene que ser siempre 3.

Si usamos símbolos de ajedrez en el ejemplo anterior, tenemos

♟ + ♝ = 4

El peón y el alfil tienen definidos los valores específicos de 1 y 3 respectivamente y no cambiarán sus valores simplemente porque el problema cambie, en otras palabras, son constantes.

En contraste, en álgebra, los estudiantes sustituirían x o y con valores diferentes cuando los valores dados hayan cambiado. En otras palabras, en ecuaciones algebraicas , los valores de x y y cambian en términos de un problema en particular. Al comparar los valores sustituidos por símbolos de ajedrez y variables algebraicas, observamos que hay una diferencia en la sustitución de los símbolos de ajedrez - es intuitiva para los niños ya que el valor de cada símbolo de ajedrez está predefinido y por ende, tiene un significado implícito intrínseco.

La idea de usar símbolos de ajedrez es enseñar a los niños a transferir un objeto concreto en un concepto abstracto. Por ejemplo, un objeto concreto (como la pieza de ajedrez alfil) puede estar asociado a un símbolo ♝, que puede ser reemplazado con un valor de 3. Este proceso de aprendizaje y destreza del pensamiento está alineada al concepto de enseñanza de Montessori.

Student's name: ________________ Assignment date: ___________________

El uso de figuras de animales u otro símbolo tal como x, y, z en rompecabezas de matemáticas y ajedrez tendrá menos significado para los niños, si lo comparamos con el uso de símbolos de ajedrez en los problemas de rompecabezas de matemáticas y ajedrez. Los niños no se confunden con el uso de símbolos de ajedrez, y les será fácil aprender las variables en la secundaria, simplemente porque aprendieron sustitución a temprana edad. Los símbolos de ajedrez son utilizados solamente como "pictogramas", i.e. como representaciones de los valores. Ellos entienden que los espacios que ocupan estos símbolos en una expresión aritmética deben haber sido en primer lugar números, cuando se utilizan en operaciones aritméticas.

La otra razón de usar símbolos de ajedrez en rompecabezas de matemáticas y ajedrez es porque los símbolos de ajedrez en sí representan los movimientos y, coincidentemente, algunas de las direcciones de los movimientos se parecen a algunas operadores aritméticos, por ejemplo la torre can la movida hacia arriba y abajo, e izquierda y derecha y así el rastro de movidas de se ve como un signo +.

Cada símbolo de ajedrez tiene una dirección de movimientos específica y estas direcciones de las movidas están "incrustadas" a cada pieza. He aprovechado la ventaja de las movidas de las piezas de ajedrez y las he definido como sigue a continuación:

Suma/Resta = Torre (también puede ser reina o rey)
Multiplicación = Alfil (también puede ser reina o rey)
División = Rey (Oposición de 2 reyes)

Valores de Ajedrez Utilizados

Los valores de los símbolos de ajedrez son los mismos a los usados en el Manual de Enseñanza publicado por la Federación de Ajedrez de Canadá (4).

La técnica de cancelación para contar los puntos de las piezas de ajedrez sería muy similar al concepto de la propiedad de la resta de un ecuación.

Student's name: ________________ Assignment date: ____________________

Los Símbolos de ajedrez y los valores son integrados en operaciones aritméticas para crear un nuevo tipo de problemas. Al usar símbolos de ajedrez mi objetivo es crear preguntas más interesantes y promover en los niños el que piensen más y utilicen razonamiento espacio temporal para resolver problemas.

A cada pieza de ajedrez se le ha sido asignado un punto diferente (valor). Los siguientes son los valores asignados a los símbolos de ajedrez.

♔ (rey) = 0 punto	♘ (caballo) = 3 puntos	♖ (la torre) = 5 puntos
♙ (peón) = 1 punto	♗ (alfil) = 3 puntos	♕ (reina) = 9 puntos

Mi experiencia de usar valores de ajedrez para enseñar operaciones aritméticas ha sido muy positiva. Los estudiantes de primaria que no hayan estudiado variables pero que si hayan trabajado con mis hojas de trabajo usando símbolos de ajedrez han absorbido el concepto de variables algebraicas o substitución en una forma natural e intuitiva. No hay necesidad de explicarles el concepto de variable sino simplemente mencionar los valores de las piezas de ajedrez, por ejemplo,

♜ + 5 = _____

Ejemplos

A continuación presentamos unos ejemplos que fui creando para mostrar las relaciones entre las matemáticas y el ajedrez.

Student's name: ________________ Assignment date: ____________________

Ejemplo 1 Suma y Resta

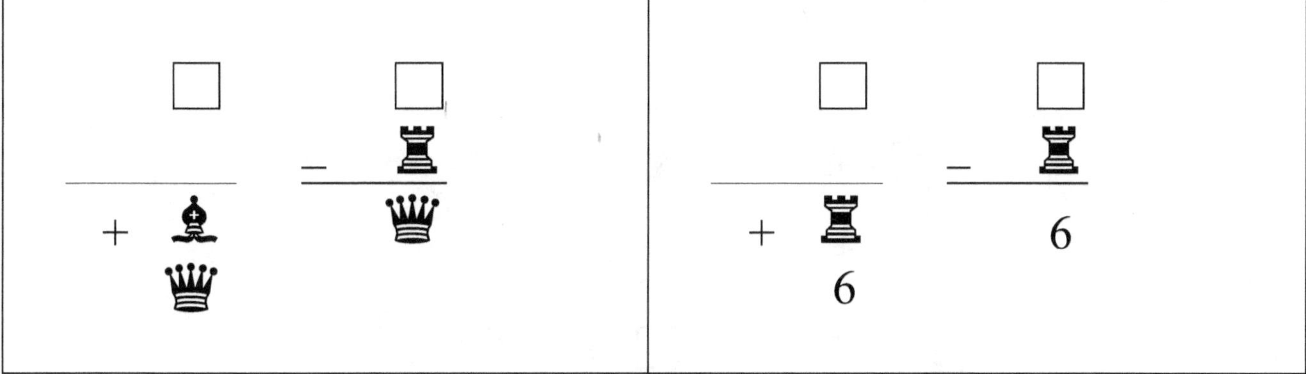

El problema anterior lo diseñé para dar una alternativa distinta a la de las hojas de trabajo convencionales, que siempre son de izquierda a derecha en forma lineal. Podríamos resolver el problema anterior desde abajo hacia arriba y luego desde arriba hacia abajo en múltiples direcciones.

Student's name: ________________ Assignment date: ____________________

Ejemplo 2 Multiplicación

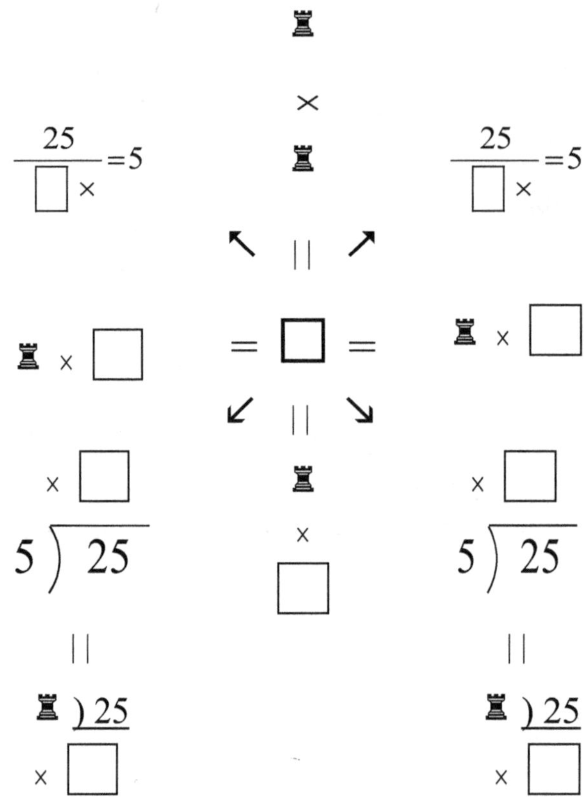

El problema anterior fue creado basado en la idea de que en la vida real, los niños no aprenden matemáticas en una forma secuencial de suma, resta, multiplicación, o división. Aquí verán cómo incorporé en la idea de la multiplicación diferentes formatos para calcular. El propósito de esta hoja de trabajo es aprender multiplicación pero exponiendo a los niños a las diferentes formas en que pueden ser presentadas las multiplicaciones. Un simple problema de multiplicación es cambiado a un problema de dos pasos multidireccional, de multi-operación, y aprendizaje multiconceptual.

Student's name: ________________ Assignment date: ________________

Ejemplo 3 Lógica

<table>
<tr><th>Símbolo de ajedrez</th><th>Ejercicio de Lógica</th></tr>
<tr><td>

Un nuevo símbolo de ajedrez está definido como sigue:

Figuras de Ajedrez	Símbolos de ajedrez
♔	÷ (Oposición)
♖	+
♘	L
♗	×
♕	✳
♙	⩛

</td><td>

En la siguiente ecuación, observe los símbolos de ajedrez en la izquierda y llene cada ◯ con un número.

Si + + + = 10

entonces ÷ + + = ◯

   + ✳
\+ + ÷
―――――
◯ ÷ ✳

</td></tr>
</table>

Use la tabla de Símbolos de ajedrez y halle el modelo siguiente:
Z, ÷, O, ⩛, T, L, T, ×, F, +, _____, ✳
Use la tabla de Símbolos de ajedrez y halle el modelo siguiente:
0, ÷, 1, ⩛, ___, L, 3, ×, 5, +, ___, ✳

El problema anterior demuestra no solo que los niños no se confundirán con los símbolos de ajedrez tradicionales que son usados en expresiones aritméticas de mis libros de trabajo, sino que también usarán adicionales símbolos de ajedrez creativos y serán capaces de resolverlos correctamente. Este problema está pensado para tercer grado en adelante.

Student's name: ________________ Assignment date: ____________________

Unos estudiantes no pudieron resolver los 2 últimos rompecabezas pero apreciaron la sofisticación de los problemas cuando les expliqué las relaciones lógicas. El último problema es el que usa una combinación de símbolos de ajedrez y sus valores. El uso de valores de ajedrez se parecen más al uso de valores monetarios. Cuando las figuras de ajedrez o las figuras monetarias son vistas por los niños, ambos representan algunos valores significativos predefinidos. El siguiente es un ejemplo donde los valores de las piezas de ajedrez pueden ser valores monetarios y "Total Puntos" pueden ser la suma de total dinero.

Ejemplo 4 Valores de la Tabla

Llene diferentes número de piezas de ajedrez para sacar un valor total.

Número de ♟	Número de ♞	Número de ♜	Total - puntos
1	1	1	9
3	2	0	9
0	3	0	9
□	□	□	10
□	□	□	10
□	□	□	11
□	□	□	12
□	□	□	13
□	□	□	14
□	□	□	15

Ejemplo 5 Ecuación

Los siguientes ejemplos muestran cómo los símbolos y valores de ajedrez son integrados con operaciones aritméticas para facilitar la destreza de pensamiento específica.

Student's name: ________________ Assignment date: ____________________

♛ + ♞ + x = 54

x = _____

Ejemplo 6 Suma y Resta, Si, Entonces - Resto...

	10	− ♚	=	□	
+	5	+ ♚	=	□	+
	□			□	

Si 10 + ♜ = □, entonces 9 + ♜ debe ser □.

Si ♜ + 10 = □, entonces ♜ + 9 debe ser □.

Student's name: ________________ Assignment date: ____________________

Ejemplo 7 Multiplicación Cruzada

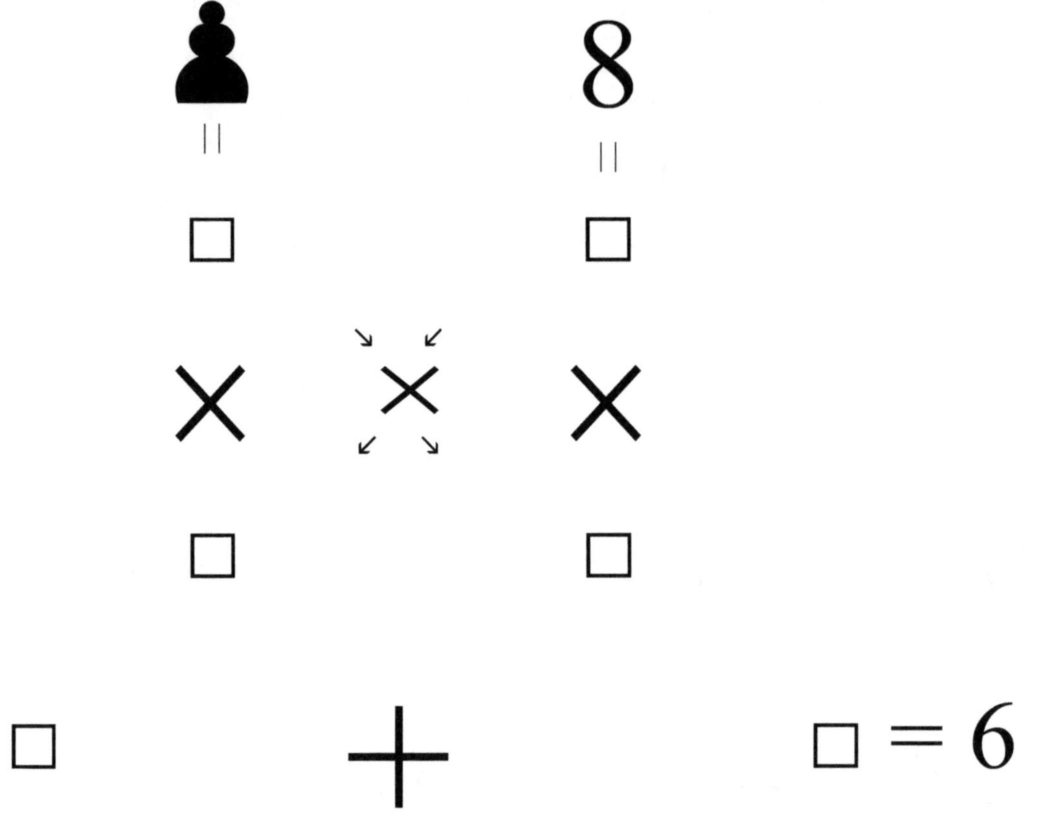

Student's name: ________________ Assignment date: ____________________

Ejemplo 8 Multiplicación y división

♛ × 2 ——— 18 ÷ 2 = □	♛ × ♜ ——— □ ÷ 5 = □
♛ × ♜ ——— □ ÷ 9 = □	♛ × ♛ ——— □ ÷ ♛ = □
♛ × ♞ ——— □ ÷ ♞ = □	♛ × ♞ ——— □ ÷ ♛ = □
♜ × ♞ ——— □ ÷ ♞ = □	♜ × 8 ——— □ ÷ ♜ = □

¿Estas operaciones causarán confusión a los niños? La respuesta es no. ♛×♞ no tendrá sentido si es explicado literalmente como "la reina por tantas veces el caballo". De cualquier modo, si es traducido a numerales -9 veces 3- entonces el producto debe ser 27 lo cual es muy lógico y los niños entienden que están trabajando con 9 × 5 y no el producto de ♛×♞.

Un tipo de pregunta similar de lógica es la siguiente: Si 2 # 3 está definido por 2 + 2 × 3 ¿Entonces 3 # 4 es? Normalmente 2 #3 no tendría sentido alguno ya que no es un operador aritmético válido pero si lo definimos claramente entonces si se vuelve operable.

Student's name: ________________ Assignment date: ____________________

Ejemplo 9 El siguiente rompecabezas requiere conocer las movidas de ajedrez.

Reemplace el □ por una pieza de ajedrez	Formas geométricas
♙	
♖	
□	
♗	

Student's name: ________________ Assignment date: ____________________

Ejemplo 10 Use movidas de ajedrez para resolver el siguiente rompecabezas.

	?		14	
?				21
		♘		
?				28
	42		?	

La primera vez que ven esto, muchos de los estudiantes no lo pueden resolver, ¿por qué? Porque están muy acostumbrados a calcular de izquierda a derecha y este ejercicio tiene que resolverse en una dirección no convencional.

Ejemplo 11
Movidas de los símbolo de ajedrez para resolver el siguiente rompecabezas.

Si 2 ♜ 3 = 5 entonces 2 ♝ 3 es = _____

Sorprendentemente, algunos de mis estudiantes no tienen problema para resolver este rompecabezas.

Student's name: ________________ Assignment date: ____________________

Ejemplo12
Movidas de los símbolos de ajedrez para resolver el siguiente rompecabezas.

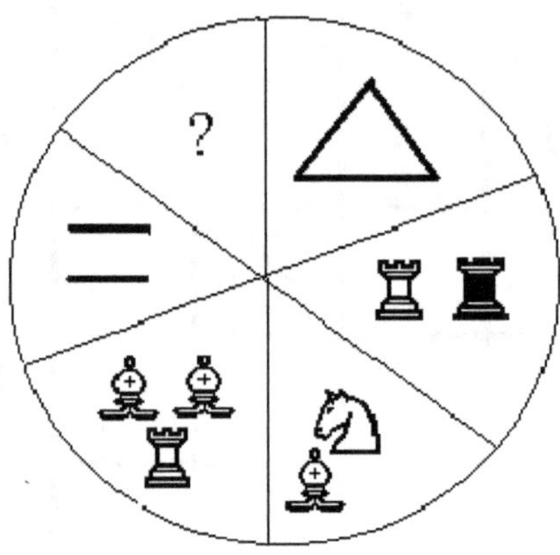

Ejemplo 13

Este problema no puede ser resuelto usando valores de ajedrez; los estudiantes lo probaron y ya lo saben. Entonces, ¿cuál es la idea que está detrás de este rompecabezas?

Student's name: _______________ Assignment date: __________________

Ejemplo 14

La Ruta de la Torre

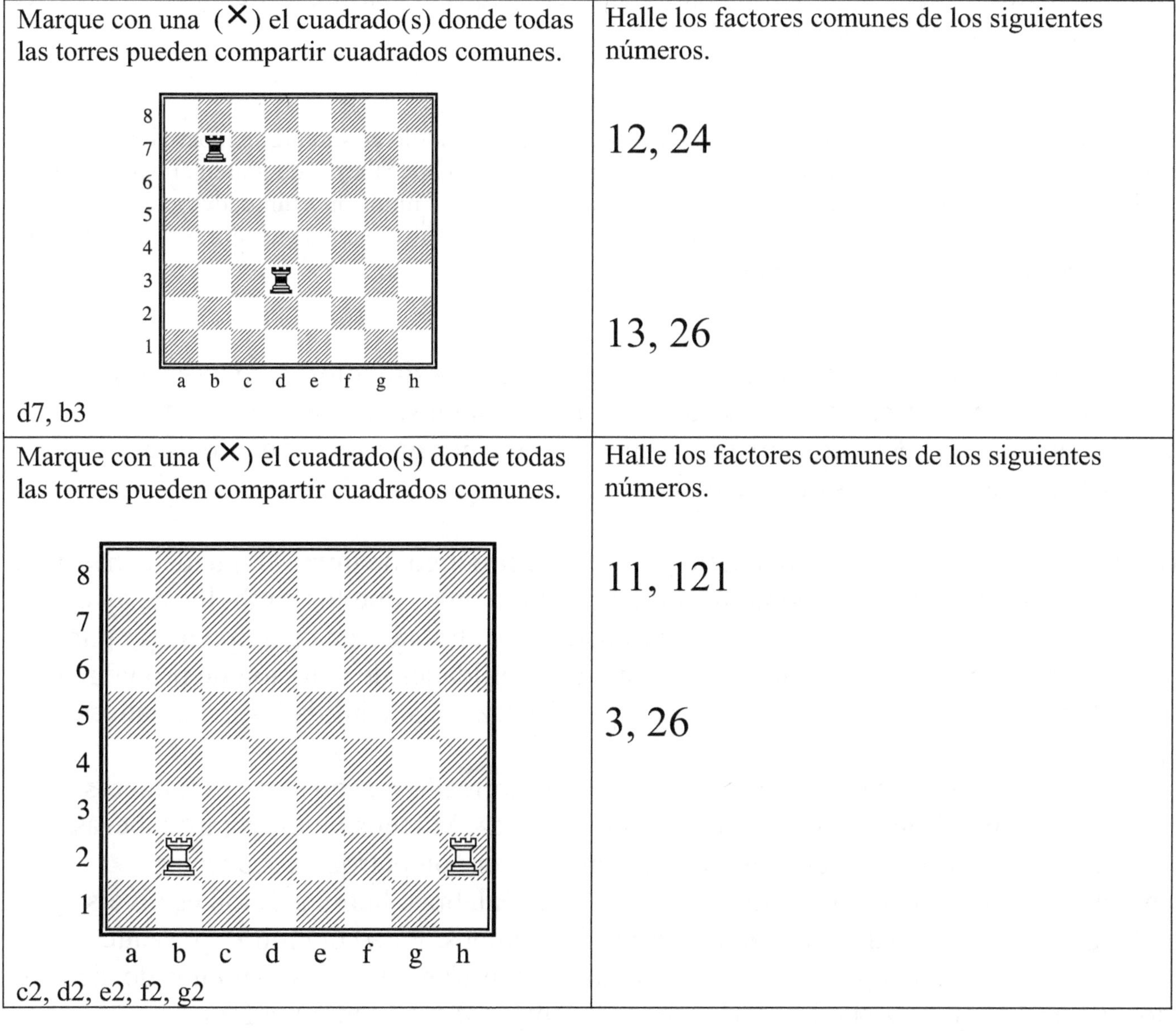

Marque con una (✕) el cuadrado(s) donde todas las torres pueden compartir cuadrados comunes.	Halle los factores comunes de los siguientes números.
d7, b3	12, 24 13, 26
Marque con una (✕) el cuadrado(s) donde todas las torres pueden compartir cuadrados comunes.	**Halle los factores comunes de los siguientes números.**
c2, d2, e2, f2, g2	11, 121 3, 26

Los problemas con movidas de ajedrez pueden ser aplicados al concepto de matemáticas, el ejemplo anterior lo coloco para demostrar la idea con un problema de ajedrez a un lado y en el otro está el problema de matemáticas.

Student's name: ________________ Assignment date: ____________________

Resumen

Encontré que la idea de usar símbolos de ajedrez es muy útil para los estudiantes de primaria ya que pueden aprenderlos para resolver rompecabezas de matemáticas y ajedrez usar su conocimiento de ajedrez.

Muchos niños no pueden jugar bien el ajedrez cuando están haciéndolo contra alguien pero en cambio se ponen muy orgullosos de ellos mismos cuando pueden resolver rompecabezas de matemáticas y ajedrez . Los Rompecabezas de matemáticas y ajedrez significan para ciertos niños oportunidades adicionales para proponerse retos. Por esta razón, yo si acostumbro a premiar no solo a los ganadores de ajedrez sino también a los que resuelven los rompecabezas.

Lo más interesante de usar símbolos de ajedrez es que los símbolos de ajedrez no solo poseen valores predefinidos sino que también implican un significado en si en sus movimientos y estas dos características especiales me permiten crear diversos e interesantes rompecabezas matemáticos con pizzazz.

Al usar símbolos de ajedrez, un simple problema aritmético de un paso, puede convertirse en uno de múltiples pasos, y como resultado de ello, vemos que los símbolos de ajedrez y los valores ofrecen a los niños más oportunidades para trabajar en otro tipo de preguntas que pueden estimular sus cerebros y mejorar sus habilidades de solución de problemas. Es un doble logro: mejorar el conocimiento de ajedrez y las habilidades matemáticas.

Los rompecabezas de matemáticas y ajedrez que he creado no son simples números reemplazados mecánicamente por símbolos de ajedrez. Muchos de los rompecabezas de matemáticas y ajedrez creados también tienen que ver con patrones, secuencia, geometría, teoría de conjuntos, lógica etc. En otras palabras, la integración es diversa y también involucra visualización en múltiples direcciones. Para concluir el presente artículo, a continuación presento ejemplos que demuestran cómo los símbolos de ajedrez y los valores de ajedrez pueden convertirse en rompecabezas con un enfoque multidireccional y multisensorial.

Student's name: ________________ Assignment date: ___________________

Ejemplo 15 Halle los valores que reemplazan a ? o llene en □.

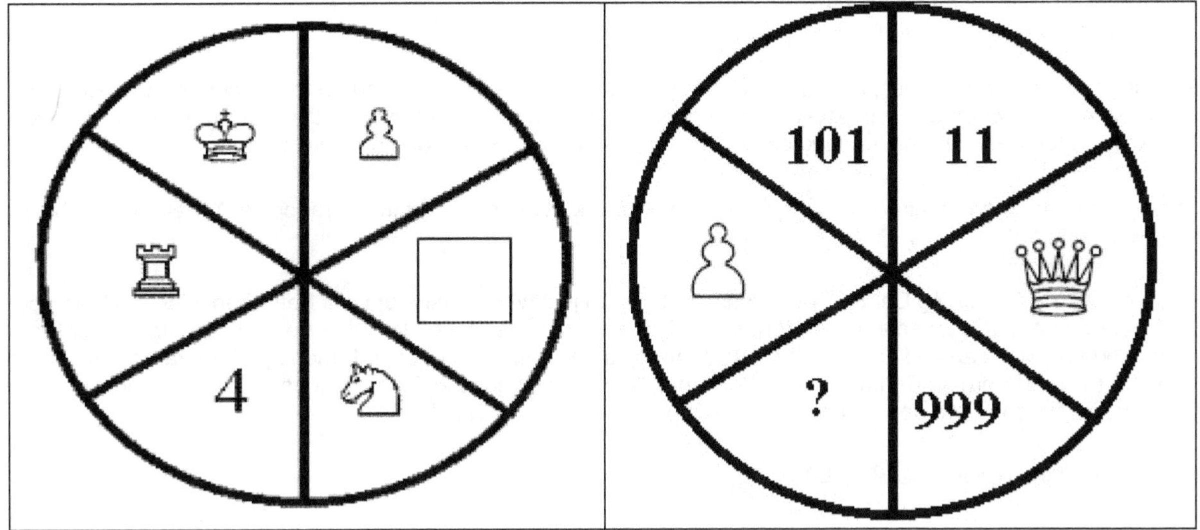

Referencias

(5) Chess in Education Research Summary, Dr. Robert Ferguson. Details see below http://www.quadcityajedrez.com/benefits_of_ajedrez.html

(6) Mathematics and Chess, by Miodrag Petkovic, Dover Publications, Inc. Mineola, New York, 1997

(7) Challenging Mathematics, Cheneliere/McGraw-Hill, Michael Lyons, Robert Lyons, 1995, Montreal, Quebec, Canada

(8) Chess Teaching Manual, IM Tom O'Donnell, the Chess Federation de Canada.

Student's name: ________________ Assignment date: ____________________

How Not to Buy a Money Burning Franchise

This article was originally published on July 20, 2005.
http://searchwarp.com/swa11788.htm

In my ten years of a tutoring business, I have seen so many learning centres come and go to Vancouver, Canada. I had a binder full of names of learning centres that were set up in the past ten years, but most of them are gone now, nowhere to be found. So if anyone tells you that the learning centre is very easy to run, could you not believe it?

It would be easier to run than running on your own is also a mistake to think by joining a franchise. My answer to this kind of thinking is caveat emptor!

There are so-called XXX Learning Center Franchises out there to just want to take upfront hefty franchise fee from you, and if you could succeed or not, it is an entirely different story. The way to make money for this kind of Learning Center Franchise is to multiply as soon as they can and milk upfront fees as much as they can out of each franchise. For example, if you pay upfront $25,000 to XXX Learning Center Franchise, then if they can convince 200 franchisees to join, then how much can they get out of the entire fiasco? Let's do the math.

$25,000 times 100 franchisees = $2,500,000

Over 2 million in a few years for the franchisor, not bad Oh? What happens to the franchisees? If the majority of them are struggling for survival but just don't have the courage to close the learning centres since each franchisee paid $25,000 for it and just could not let it go? It is a sad story.

Do you like the landlord who gives a 20% rent increase when it comes to renewal time with no justification? Unexpected and sad surprise for the business owner. Who needs this kind of life?

Don't burn your money by buying a bad learning centre franchise. Buy a good and reputable one with a very reasonable franchise fee.

Many learning centre franchises give you the "big" talk and mumble and jumble about how great their material or teaching method is. If you examine carefully, you will find most of their description of how wonderful their material has no real substance.

How is their math material different from others? Or is it different from the textbooks used at school? Is it different only in quantity? Are the questions only harder? Why the material increases children's ability in thinking skills? Are there any research papers done on the kind of materials been promoted?

Most of the math learning centres could not answer the above questions clearly or will elude the question by answering something like our material is great and our students like them etc., without touching the real issue that is how the material is unique?

You got to know what product you are buying after spending $25,000 upfront fee. This is the most important discriminating factor in buying a math learning centre franchise.

If the founder of the learning centre did not have any specific philosophy on how math should be taught when setting up a learning centre many years ago then how could we expect him/her to have one 10 years later? It is a joke to read their statement on why their math material is good and useful for children when they did not have any clue in setting up their first math learning centre.

Why buy a Ho Math Chess Franchise?

Intellectual Properties of Ho Math Chess

何数棋谜 - **www.homathchess.com**

Student's name: ________________ Assignment date: ___________________

Ho Math Chess Learning Centre started with the view that math should be taught in a multi-sensory way, that is to not just do computations. There are hundreds of research reports on learning chess benefits kids in many different ways. The only problem is there were no materials that integrate math and chess together so kids could learn math and chess all in one until I created them and started to use them at Ho Math Chess learning centre.

We have the world. First, one of its kind, very unique, and copyright protected math product for grade 1 to grade 7. You can examine my math products on the web right in front of your eyes. Go to www.mathandchess.com and download a few samples of my worksheets. Take a look.

The math learning centre business is very competitive, and only the best will survive in the years to come, so don't buy into a math franchise with no unique product. Buy a math learning centre with a unique, marketable and effective product.

You should not just buy the franchise name. The product is equally important.

Student's name: ________________ Assignment date: ____________________

How to Buy a Franchised Math Learning Center

The era of using the idea of "cut and paste" some math worksheets together to open a math learning centre has long gone. Children today living in a world that has a landslide difference from their parents, urban children do not run around in the rice fields and they hardly play with their neighbourhood children anymore. Many children grow up surrounded by computer games and cell phones and downloaded music and they watch video clips on their cell phones. Simply put, it is an electronic age and video game world they grow up with. With the above scenery in mind, then we look at what kinds of math worksheets today's schoolchildren are working on. Lots of children are still working on an old style of math drills, and they dislike them.

Why most children hate old-style drill math worksheets? They are downright boring, lifeless, and not interesting. Computer-based teaching suffers the same fate as dull worksheets. There is no bonding between children and computers. Even to use the point that children learn by mimic their mom's language to justify math drills, we still cannot deny the fact that mom has different facial expressions, gestures, different tones and the most important, mom has an emotional feeling toward their children, and for these worksheets and computer offer none of them. This is why a human math tutor cannot be "easily" replaced by the use of drill worksheets or a computer-based teaching method.

The fact of the matter is also we need children to continue to practice to get the fluency of computation procedure to work out problems, and for this, we still need math worksheets.

The far and foremost important investigation to buy a math learning centre is to take a look at what kind of worksheets children will be working on. The worksheets are tied to the teaching philosophy or mission statement of the math learning centre. For example, Ho Math Chess started because its founder has a personal story to tell, and the founder believes children learn best while having fun. It does not matter how wonderful your math tutors are, but if children do not want to enrol or are not motivated to learn, then the results will not be there.

Student's name: ________________ Assignment date: ____________________

What is wrong with the old style of math worksheets? Many worksheets have been produced by math educators but have these worksheets gone through the stages of been actually "sampled" by children to find out what they like or dislike? Children like to be challenged by not under stressful conditions and children like puzzles and like to use their brains to think but not to the extent that they get the feeling of hopelessness. Children enjoy that problems can be solved by themselves step by step if they put a bit of effort into trying and are able to think for themselves. The difficulty is how we, as math educators, are going to create some math worksheets to possess these characteristics?

A good math worksheet has a clear goal, has a message to convey to children by the creator and children are interesting in hunting for the solution by going through some kind of logic process. There are very minimum efforts required to explain the question, and even though the length of thinking maybe long, children do not even realize that. Good math worksheets are teasing, interesting, and addictive. The difficulty level can be low or can be high, and there is a uniform style across all worksheets with a theme. These kinds of worksheets are good to train children's basic math and motivate children to learn. A good example is Franko ChessDoku, which combines Sudoku, math and chess all in one. Children learn logic, chess and improve math at the same time and all on one worksheet.

Children like mazes. They are adventuresome and interesting. It gives children a feeling of reward when the problem is solved, and children feel better when they are challenged mentally. The difficulty is how to marry math and mazes? The problem has been solved by integrating math into chess mazes. This is the birth of Frankho ChessMaze. A chess maze is not only fun to play, but it also teaches geometry concepts and trains children's visualization and orientation.

Some children were born with sensitive number sense, but most children can be trained to be more alert with numbers, patterns, and symbols. Chess serves as a good tool to combine with math so as to provide an innovative way of training children to be more aware of number patterns and the relationship between numbers.

Student's name: ________________ Assignment date: ____________________

By use the technology of Geometry Chess Symbols invented by Mr. Frank Ho, children are able to work on an array of different math worksheets, which are to be more entertaining, more fun and more interesting.

For more details on how math and chess integrated worksheets can raise children's math marks with fun, visit www.homathchess.com.

It is my view that a good math learning centre has the potential to become a worldwide franchised math learning centre by firstly producing a one-of-its-kind and innovative worksheets which children truly like and enjoy working on them. Secondly, these worksheets must be streamlined and have a uniform look to have the potential to become a popular product. Because of these extremely unconventional worksheets, a business model can be created to also streamline the teaching flowchart to provide an efficient and smooth teaching method.

No success can be achieved if the unique math worksheet, which is the key ingredient of franchising a math learning centre, does not exist. Not only are these worksheets truly interesting to children, but they also can improve children's math ability. Otherwise, they are just fad and will not have a long-lasting effect.

Good worksheets motivate children to learn, and at the same time, they also improve children's true math ability.

When buying a franchised math learning centre, one should look for the true substance to see what a math learning centre can really offer, whereas others cannot.

Student's name: _______________ Assignment date: _________________

How to Buy a Specialty Math Learning Center Franchise

The future trend of the learning centre industry is to go specialty. The competition is so fierce among learning centres; newcomers would have a hard time to survive if the after-school learning centre is not specialized, much like many customers now go to the shoe store to only buy shoes now. To buy a math specialty learning centre, the first important thing to do is to look around and find out what types of math learning centres are available.

The most traditional way of teaching math is to use a drilling system. Although this type of learning center is much against by the modern math scholars, they are many still around, especially in North America due to the reason the homework is not as heavy when compared the other parts of the world, especially in Asian countries. So many of these types of "drill" learning centres are called themselves self-learning centres since they virtually do not teach kids math or help school math work. They are not doing well in Asia. Why? Asian school teachers already assign so much math homework to students; students do not really need more exercises to practice. What they really need is to get more "teaching" time, not "practice' time. So the future of this type of learning centres is very limited now especially more and more parents turn to another style of learning centres when kids are complaining how boring of these self-learning teaching methods in the disguise of just pure "drills". Ironically, they are effective in improving computational skills but are downright boring. If parents believe math is more than just computation, then this teaching method is pretty much outdated. It has a franchise advantage that is it so closely follows the successful principle of a franchise that is "teaching" is just to mark homework – a repetitive task. From franchisee to franchisee, the teaching really relies on their teaching material, not teachers. Thus the concept of fast food franchise idea can be found here. Students go through a routine of having homework marked or checked, and that is about all these learning centres will do.

Student's name: ______________ Assignment date: ________________

The teaching method also encourages parent's involvement in actually teaching their own kids, but quite often, it gets into arguments between parents and children and it hurts each other's feelings. This is the drawback of the system. On the surface, ideologically, the teaching method sounds very good, but in reality, it has problems. Since it advocates low fees and uses the fast food concept to the franchise, it attracts many copycats to use more or less the same "drill" method but with modified formats.

There are also learning centres with lecture-style that is the learning centre will teach students regardless what the students are doing at school, it is fine if the students do not have problems at school math but if they do then this type of learning centre can only attract the brightest kids to go. They really have no uniform teaching material, so it is difficult to franchise, and many independent math learning centres use this type of teaching and managing style. Once the superstar teachers are gone, then the class also dies since the teaching almost exclusively relies on the star teachers to attract students. There are no real teaching methods, no real teaching philosophy, and no systematic workbooks. It has the drawback of selling as an asset in the future since there is really no asset to sell to others other than the superstar teachers. A successful franchised math learning centre must weigh heavily on the teaching workbooks, and they must be unique with innovative teaching ideas.

There are some other new and innovative math learning centres like using an abacus or some type of Indian mental computing technique, but they also have their shares of problems. The biggest problem is their math computation techniques are so different from the way most schools are teaching, and most kids abandoned what they learned as soon as they do not register at these learning centres. Without continuing to practice what they learned, they will not use these techniques when going to their day schools. The benefits of learning these so-called "effective and innovative" ways of computing go wasted if they go against the mainstream teaching method and technique.

Student's name: ________________ Assignment date: ___________________

Finally, there is another type of learning centre that does not use any out of the world computation technique that deviates from the mainstream technique but tries to come with a different teaching method that is more effective or fun. Many in the past have tried this method but failed since no one could successfully come up with a workbook that integrates game and math, making the learning environment both fun and educational. This has all changed when Ho Math Chess, founded by a Canadian math teacher Frank Ho, invented the teaching method by combining chess and math. It does not do math teaching and chess teaching separately. Ho Math Chess created the world's first commercially available math, and chess integrated workbooks and successfully implement this teaching method in the worldwide Ho Math Chess Learning Centers. It solved the problem that has boggled educators for so long: how to create a learning environment for kids such that children could learn math while having fun?

This "educatmaint" teaching method will be the future for the math specialty learning centre. With the integrated math and chess innovative teaching method, Ho Math Chess has revolutionized the entire math learning centre industry and bring the birth of a new age to after-school math tutoring and teaching. More details can be found at www.homathchess.com.

Student's name: ________________ Assignment date: ____________________

How to Find a Learning Center Franchise Using Good Math Worksheets

Math is not about just teaching computation, nor is the most important result to get awards for math contests or how high marks one could get on SAT even though these tangible results do serve short term-goal in one's life. From my view as a math teacher with over 20 years of teaching from kindergarten to grade 12, I see there is one more important long-term goal for doing math reasonably well, that is to train their creativity skill for their future.

So how to choose the right math learning centre which trains children's thinking mind? The first step is to look at what is the learning centre's teaching philosophy? It is not enough to only understanding the teaching philosophy. One must check to see if they do what they say? What happens if all the learning centre does is just drill children or give them assembled materials with very little training for thinking skills?

What is wrong with the traditional math worksheets? What is wrong is the style of worksheets, which have not progressed at pace with technology advancement. We see many toys made today are so different from, say, just a few years ago with new technologies such as sound, screen, transformations capabilities, etc. But how about the math worksheets children are working on today? Not much has changed in terms of the computation sheets. They are still dull and boring. Today, children are getting all kinds of information from different media. Children are multi-tasking on many "toys," which are multi-function and multi-operation, so do our computation worksheets reflect the life our children are living in now? These worksheets are out of date simply because they have not caught up with the change of society. Children are downloading/uploading, viewing, instant messaging, playing games, and even calculating – all these are converted into one small device. But our old worksheets only do one function: to train how to add, subtract, multiply or divide, etc., for primary students. So why these children shall feel happy doing these dreaded worksheets? In reality, children today, even as young as kindergarten to grade 4, don't just process information using only four operations: addition, subtraction, multiplication, or division.

Student's name: ________________ Assignment date: ___________________

They process information using images or numbers, and they need to sort, classify, match, compare, compute, and browse from one web address to another to complete their tasks. Children are used to multi-function and multi-task, multi-concept type of information.

Is there such a thing that math worksheets can be created to be fun-oriented that children can play with numbers and be rewarded with satisfying and great fun? Ho Math Chess has created a series of math and chess integrated workbooks to put math, fun, thinking, creativity, and memory improvement altogether. Ho Math Chess uses international chess as a tool to converge the information of images, patterns, sorting, comparison, matching, tables, etc., all together on computation worksheets to better reflect what children are facing or already doing today. Ho Math Chess has successfully created the world's first math and chess integrated workbook using its proprietary intellectual product Geometry Chess Symbol. The astonishing result is children are more willing to work on Ho Math Chess worksheets than traditional worksheets. They are more engaged in the Ho Math Chess activities when compared to working on the traditional worksheets. With Ho Math Chess worksheets, children learn math in a fun way.

It is crucial not only to look at the teaching philosophy of a learning centre but also to determine if the learning centre has worksheets to back up what it is claimed. One should ask this question: Do the worksheets speak for the learning centre's teaching philosophy?

More information on How to Find a Learning Center Franchise Using Good Math Worksheets can be found at www.mathandchess.com.

Student's name: ________________ Assignment date: ____________________

How to run a successful franchised math learning centre

How do we judge if a math-learning centre is thriving? It depends on what type of learning centres. There are many types of math learning centres. Some learning centres mainly prepare children for tests; some are for comprehensive skills training, including thinking skills. Does it depend on what parents want? Some parents only look at if their children's marks have improved, while some parents also like to see if the math program can improve their children's thinking skills. To put it in a nutshell, a thriving learning centre is where children like to go and also enjoy it.

A thriving math learning centre is a place where children are smiling and happy while learning at the centre and then come home telling their parents what they have learned and are happy to go again. A successful math-learning centre makes kids learn in a happy mood.

A successful math-learning centre will not be successful if their materials are boring and inhuman. Why will children be happy if all they do at the learning centre is drilling on math worksheets, even it means their computation ability has improved? The untold negative side effect is they will not have fond memory and experience with this kind of learning centres. My students in my class tell me that they "hate" this kind of math learning centre, but some of their parents "forced" them to go. I can say these unimpressed children will not send their kids to this kind of learning centre again when growing up to be parents one day.

Young children are happier to go to a math learning centre, where they do not just sit there and do pure math work. The problem is most math learning centres cannot offer what these young children want. I know what kind of worksheets children like to work on since I have offered them choices, and I have the chance to observe their reactions and feedbacks.

Children like to be entertained while learning; learning while having fun is an excellent way for children to learn.

We cannot turn the math learning centre into a circus, so children have lots of fun but do very little math. This is not an idea of integrating fun and math learning.

Why buy a Ho Math Chess Franchise?

Student's name: ________________ Assignment date: ___________________

Many children like puzzles but not math work, so how do we "trick" them to do puzzles while they also have to do math work? We are not talking about computer games or computer-based learning here. We also are not talking about giving children some separate puzzle worksheets in a math class. We are talking about a pencil, paper and re-printable "game, puzzles, and math" integrated worksheets.

If a math learning centre can integrate math and game such as chess and puzzles all into one worksheet, then this is what children prefer to do rather than the traditional type of math worksheets.

How to deliver these fun worksheets systematically so that they will also increase children's math marks is very important to make a math-learning centre succeed.

A successful math-learning centre must have fun worksheets, and these fun worksheets also foster the environment of thinking.

A successful learning math-learning centre offers something that children like, and the results are children are happy to learn, and their math marks also improve while having fun.

Student's name: ________________ Assignment date: ____________________

How to Run Math Franchise Successfully

The other day, someone suggested that Ho Math Chess also offer an SAT course, so I asked why? I was told that there is a demand, and also many learning centres now offer SAT courses. The reason for Ho Math Chess also jumps on the bandwagon seems to be convincing, but I would prefer to look at this issue in a long-term view, and this prompts me to spend the next few days to ponder on a topic that is just how to run a math franchised centre and be successful? If I know the answer correctly, it will also answer the perplexing question if Ho Math Chess should offer SAT and many other similar questions?

A franchised math-learning centre will never make it to the international stage if it just acts as a follower. Ho Math Chess continually evolves and innovates. Many new workbooks were created because we know they are more efficient and help children improve math, much like pharmaceutical companies continuously looking for new and improved drugs. It is cumbersome that students have to work on different sets of worksheets to get combined effects of improving math, fun with puzzles and brainpower improvement, so Ho Math Chess invented multi-function math worksheets to not only improve student's math ability but also improve their problem-solving ability and all these are done in a fun environment.

A franchised math-learning centre must have a clear teaching philosophy. All workbooks and teaching programs are developed based on its philosophy. Otherwise, the franchised learning centre may just become a tutoring centre for marks, and that is all. With this kind of narrow pedagogical view, this kind of franchised math learning centre will not have uniqueness and one-of-its-kind teaching materials. Why Ho Math Chess has to offer something just because others are offering? Ho Math Chess stands out above the crowd because of its uniqueness in recognizing a niche in tutoring and its most effective workbooks.

Student's name: ________________ Assignment date: ____________________

It takes time and effort to develop teaching materials and develop a standardized teaching system so that the right franchised math learning centre will not just hastily jump on a hot product for a short-term gain. What happens if the SAT is no longer required for university entrance requirements?

A good and successful franchised math-learning centre must be unique and be different from others in its teaching philosophy and reflected its philosophy in its student workbooks or teaching materials. Students will increase math marks under guidance. They must also enjoy the program and feel it is really fun to work with, which cannot be achieved by having a snappy marketing slogan. This is why Ho Math Chess spends a large proportion of its time and efforts in writing workbooks and creating programs for a long-term view. Many sustainable products like math, chess and puzzles integrated workbooks, Frankho ChessDoku, Frankho ChessMaze, Amandaho Moving Dots Puzzle, and Mathematical IQ puzzles are all created for a long-term view.

A franchised math learning centre's brand name is most effective when students and parents can identify it through its unique teaching method and one-of-its-kind workbooks. A successful franchised math learning must be a leader and not a follower in creating innovative products and have a standardized teaching method.

Student's name: _______________ Assignment date: __________________

How to spot a good children's education franchise opportunity

Children's education franchise comes in different forms. Some are in physical training, such as sports, some involve in art, and some help children develop their academic abilities such as English or math etc. If you love to be with great patient children, then operating an education franchise may be the perfect business for you. Still, of course, interest alone will not make the business viable. A factor to be considered when buying an education franchise is the uniqueness of the franchise system. There are many competitors out in the world of after-school math tutoring centres, so how to pick one, which has the best success rate? The uniqueness of the franchise will make a difference.

A math tutoring franchise centre is one of the best children's education franchises since all children in all countries must learn math from kindergarten to grade 12 as one of the compulsory subjects. Many children need extra help in trying to understand math.

Technology changes, parents' high demand, and fierce competition have caused sea changes in the math franchise education system. Traditional math drill worksheets are signs of an out-dated teaching style. Children raised in the multi-media environment are no longer happy with drill-style learning, no matter how one would package it in different slogans such as self-learning etc. methods. A computer-guided learning method invented a long time ago would have replaced all teachers if children accepted it. Today's children want to have math worksheets designed with their thinking in mind that are fun, exploratory, and educational. Yet, educators have not come up with innovative worksheets to reach the stage of "you order, we deliver". In other words, child education franchisers are not at the stage of delivering products to children and effectively raising children's math ability.

It is not enough to buy a math-learning centre just because the franchiser claims the system is the most effective. Why? What factors have made the system the most effective? If the math computation technique method is not what children would learn in their day school, how long will this non-mainstream computation method remain in children's memory if children do not practice them often? I have met so many children who abandoned some computation techniques they learned at other learning centres just because they did not have a chance to practice after learning. It may impress parents when children learn this particular method, but what happens after finishing the program? They go back to the pencil and paper method.

Student's name: ________________ Assignment date: ___________________

Math learning centres must use a computation technique in line with what children are learning at school, so it complements children's learning instead of learning some new techniques. The consequence of the learning is that they add more burdens on children.

If children already fear math or feel it is no fun to do math work, what good is it to add more worksheets and have them do more work? Most math learning centres fail because they ask children to do more what they feel dreadful. The drilling method is even worse since the teaching and understanding concepts are not considered important other than repetitions of practice.

It is challenging to develop a system that can add spice to math skills practice so children will feel it is fun to work with and are more willing to work on it. The main reason is that no one has ever come up with practical math workbooks on how to systematically mix fun and math. Ho Math Chess is the only world child education franchise with successfully mixed chess, math, and puzzles, so children will feel it is more fun to work on than traditional math worksheets.

To integrate math, chess, and puzzles altogether, we need innovative ideas to do it. Ho math Chess has invented the technology to do just that. The technology is a copyrighted and trademarked proprietary intellectual product.

More details about Ho Math Chess can be found at www.mathandchess.com.

Student's name: ________________ Assignment date: ____________________

Why can Ho Math Chess workbooks help students?

I was curious to find out why a child already in grade 5 could not do basic multiplication well despite how I explained to him concepts. I was equally at a loss to find out why a grade 10 student who could only "see" how a problem can be done after I explained three times or write the solutions step by step and allowed him to stop me whenever it was necessary. In this case, I know the grade-10 student never 100% understood the math instruction in his day school because he could not stop his teacher's instructions just for him on every computation step. For example, When factoring (x square – y square), the student just could not see how this problem can be done when the problem changes to (x + 1) square – (y+1) square.

After I did my observations and thinking, I found out these problems have lots to do with their math training while they were young other than their ability deficiency in basic math. They lack training in the following areas:

1. Pattern: They have some difficulty in seeing patterns.
2. Visualization: They have difficulty to see the relationship between columns or rows or Table to table or from numbers to graphics and vice versa.
3. Abstract: They seem to have difficulty to grasp the abstract idea. Any problem requires students to do a conversion using abstract ideas, then they seem to run into problems.

To take care of these above-mentioned problems, Ho Math Chess has created a series of workbooks to train children in patterns, symbols, and relationships using visualization.

The way we do it is to use our copyright protected invention Geometry Chess Symbols to have created a series of one-of-its-kind workbooks such as IQ puzzles, Frankho ChessMaze, IQ mathematical chess puzzles, word problems, and math contests to strongly increase their brainpower in number intelligence.

More information on how to improve children's math ability can be found at www.mathandchess.com.

Student's name: ________________ Assignment date: ____________________

What is Frankho ChessDoku?

The following is a sample of Frankho ChessDoku.

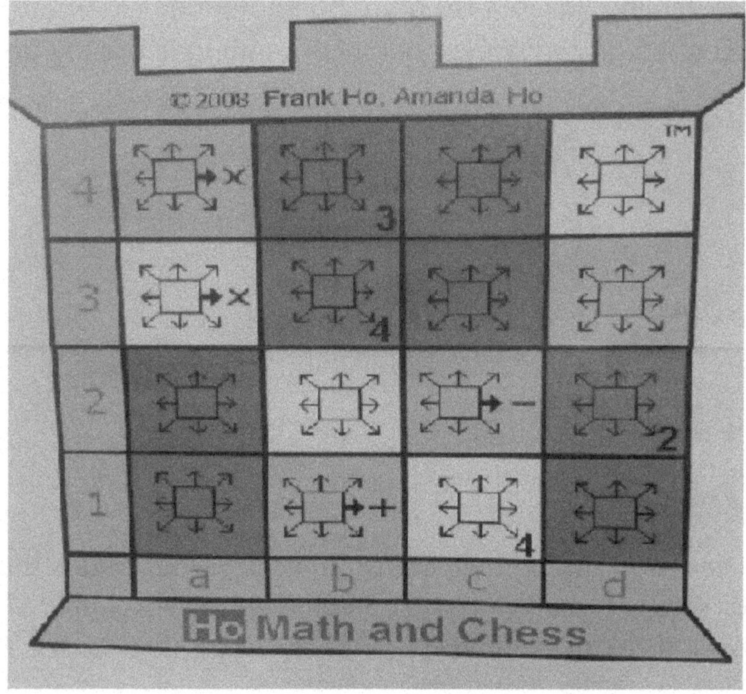

Frankho ChessDoku was invented by a Canadian math teacher Frank Ho (1, 2). Seeing the popularity of Sudoku but with no computation capability, Frank decided to do something about it, so Frank used his invented Geometry Chess Symbols (Canada Trademark TMA771400, copyright 1069744) along with Sudoku created the *Frankho ChessDoku* in 2008. *Frankho ChessDoku* is a unique puzzle that combines chess and Sudoku and is specially designed for children to solve arithmetic Sudoku by following chess moves. In 2009 Frank Ho and his wife Amanda Ho published a *FrankHo Math Chess Puzzles for Children* workbook. This workbook is now available from www.amazon.com.

Frank always has an idea about teaching math: students should always be encouraged and given a chance to THINK, and it means even when they are doing pure computation problems. This is why he has created many computation workbooks without obvious problems presented to children. Instead, children have to figure out what to do by going through a puzzle-like thinking process. *Chess + Sudoku = Fun Frankho ChessDoku*

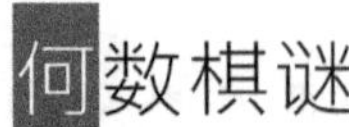

Student's name: ________________ Assignment date: ____________________

The pleasure of working on these kinds of workbooks could be very well described by a famous classical Chinese poem 山重水复疑无路,柳暗花明又一村 (Equivalent English phrase is *seeing the light at the end of the tunnel.*)

Frank has described the feeling, in Chinese rhyming sentences (打油诗), when working on our math, chess, and puzzles-integrated workbooks. Its meaning is mainly to describe the miracle of puzzles.

只见棋谜不见题 劝君迷路不哭涕
数学象棋加谜题 健脑思维眞神奇

Introduction of CalcuDoku

The original CalcuDoku was invented in 2004 by a Japanese teacher Tetsuya Miyamoto in Japan (3).

Comparisons

The critical difference between *Frankho ChessDoku* and CalcuDoku is that *Frankho* ChessDdoku uses Frank's invented Geometry Chess Symbols to guide children on the directions of arithmetic operations instead of using "boxes or "cages" as used in Miyamoto's puzzles.

Frankho ChessDoku does not just use chess pieces to replace numbers in Sudoku, as seen in some ChessDoku puzzles. *Frankho ChessDoku* invites children to trace chess moves to see the results just as if they were playing a chess game by examining the intersections of chess moves and then use the logic of Sudoku to figure out the answers. Both strategies of playing

Student's name: ________________ Assignment date: ___________________

a chess game, especially chess moves and the arithmetic Sudoku logic, needs to be combined to solve *Frankho ChessDoku* puzzles.

Miyamoto runs a learning centre in Japan and teaches his puzzles to children. Frank and his wife also use their puzzles to teach children from age four and up in their learning centre in Vancouver, Canada. Both Frank and his wife teach kindergarten to grade 12 math, and both of them also teach math contest preparations.

From a student's learning math point of view, *Frankho ChessDoku* offers more robust learning and mental training advantages over regular Sudoku and other arithmetic Sudoku types. The following table gives some comparisons. In addition to being a fun puzzle, *Frankho ChessDoku* is more suitable for students who like to improve their brainpower and mental math ability.

	Frankho ChessDoku	Regular Sudoku	CalcuDoku
Plus, minus, multiplication, and division	Can provide four mixed operations by following chess moves within one equation with no confusion.	No computations	Only independent and separate +,−,×, ÷ operations can be provided. Mixed four basic operations could confuse young children.
Vertical, horizontal, and diagonal operations	The horizontal or vertical operations are provided. The diagonal operations can be provided. The "jump" operation (knight move) can be provided.	No computations	Only horizontal or vertical operations are provided. No diagonal operations can be provided. No "jump" operation can be provided.
Framed or boxed operations	No framed operation is required since chess moves guide the operation direction. Children can always circle operation statements themselves. This flexibility allows intersecting "boxes" with no confusion.	No computations	Since the operation is always "boxed" or "caged" with a single operation, no flexibility is allowed for mixed operations. Intersecting frames or boxes would cause confusion.

Student's name: ________________ Assignment date: ____________________

Example 1

Rule

All the digits 1 to 3 must appear exactly once in every row and column. The number that appears in the bottom right-hand corner is the result calculated according to the arithmetic operator(s) and chess move(s) as indicated by the darker arrow(s).

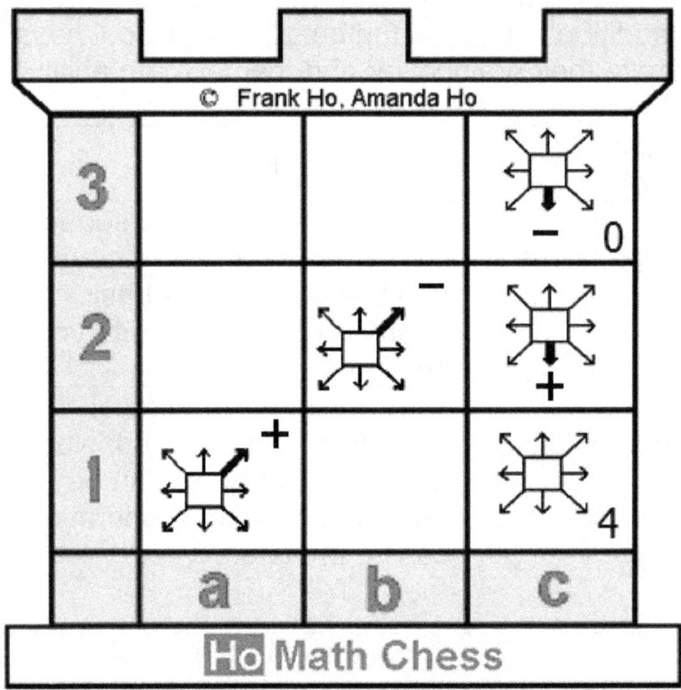

Step 1:

Circle all operations by following chess moves.

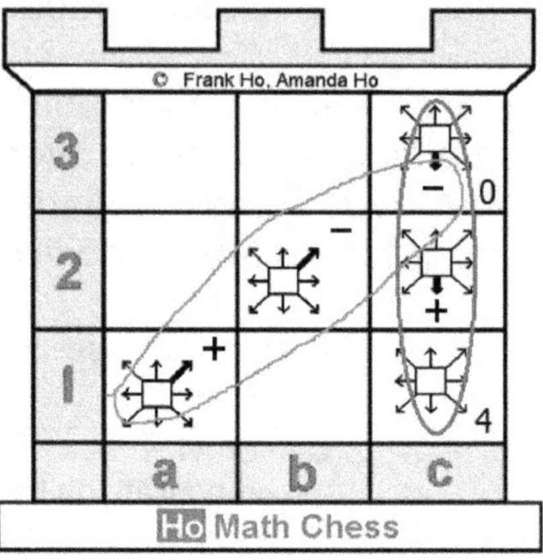

Step 2.

For the diagonal oval, 2 + 1–3=0,
1 + 2–3=0.

For the vertical oval, 3 – 1+ 2 = 4
So we know c3=3.

The final answer is as follows.

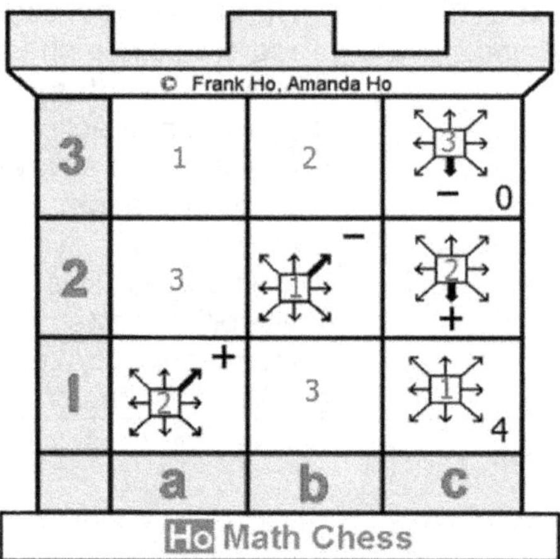

Student's name: _______________ Assignment date: __________________

A CalcuDoku is not able to produce the diagonal operation of the above *Frankho ChessDoku*. The mixed operation 3 + 2−1 = 4 or 2 + 3−1 = 4 is also difficult for children to work on if it happens in the CalcuDoku since it involves two operations simultaneously but is very easy for *Frankho ChessDoku* to identify it with no confusion. This deficiency in CalcuDoku means children seem to be stuck with only one operation at a time with very little chance to work on mixed operations. On the other hand, Children working on *Frankho ChessDoku* will have plenty of opportunities to work on either single or mixed operations with no confusion only by following chess moves.

Student's name: ________________ Assignment date: ____________________

Example 2

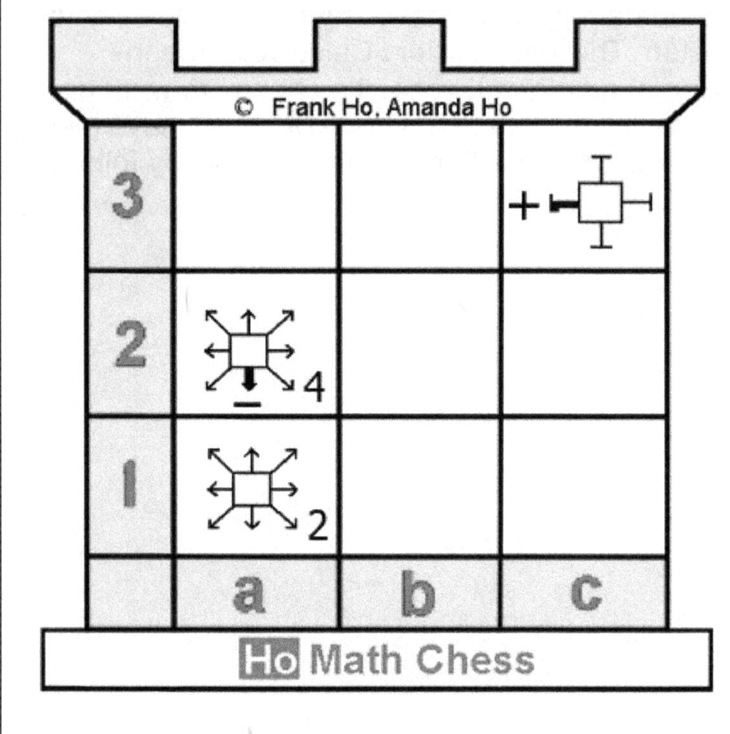

Step 1

Circle all operations.

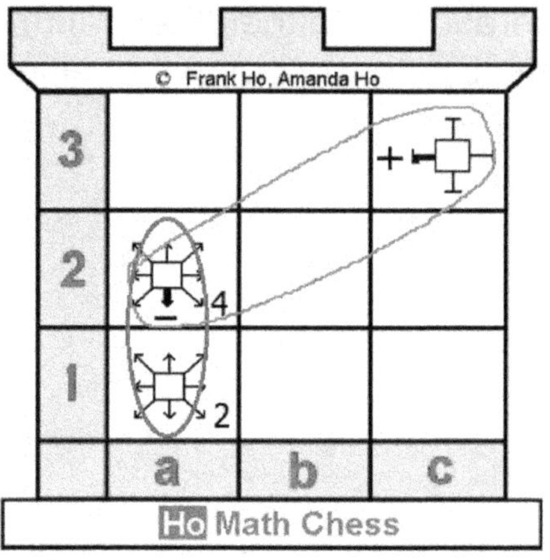

Step 2

For the diagonal oval, 1 + 3 = 4, 3+1 = 4, 2 + 2 = 4.

For the vertical oval, 3 – 1 = 2. So we know a2 = 3.

The final answer is

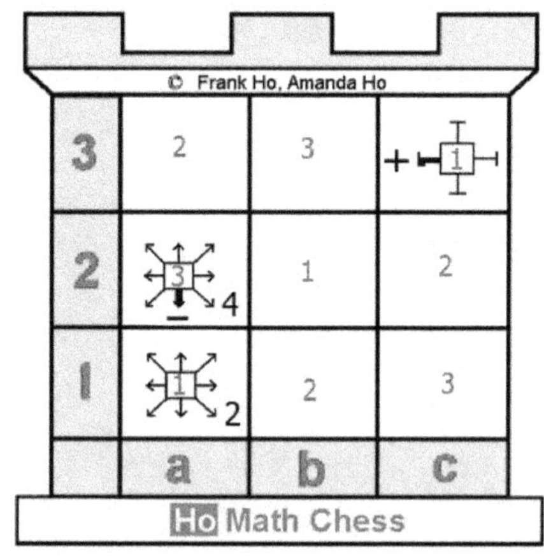

A CalcuDoku cannot produce the above "Jump" movement as acting by the chess knight move at c3.

Student's name: ________________ Assignment date: ____________________

Example 3

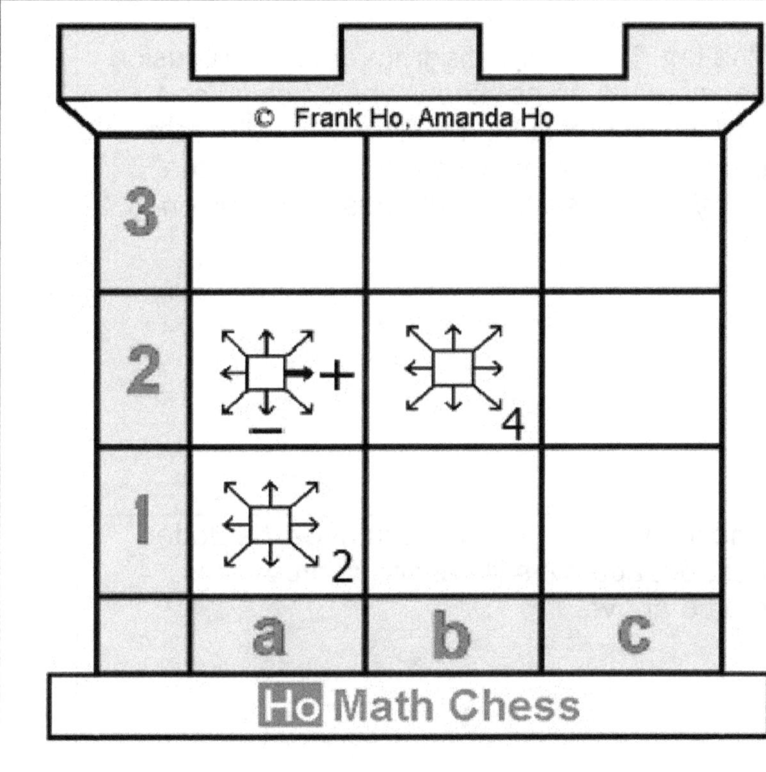

Step 1

Circle all operations.

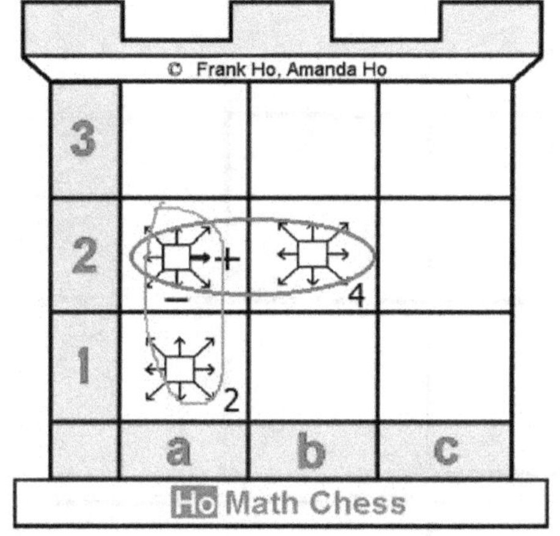

Step 2

Start at intersection a2.
For the horizontal oval, 3 + 1 = 4, 1+ 3 = 4, 2 + 2 = 4.
For the vertical oval, 3 – 1 = 2
So we know a2 = 3.

The final answer is as follows:

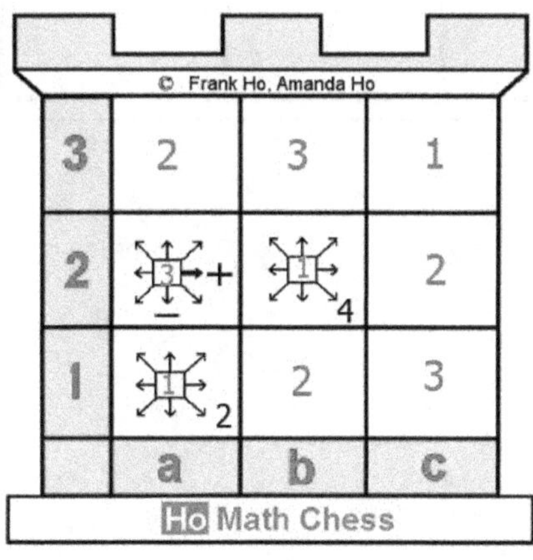

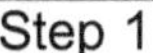

This world-famous Sum and Difference problem can be easily illustrated by using the above *Frankho ChessDoku* diagram with intersection and can be very easily solved, but trying to create it using the idea of CalcuDoku demonstrates confusion for children.

Student's name: ________________ Assignment date: ____________________

The following are the same problem (Sum and Difference) using the diagrams of CalcuDoku.

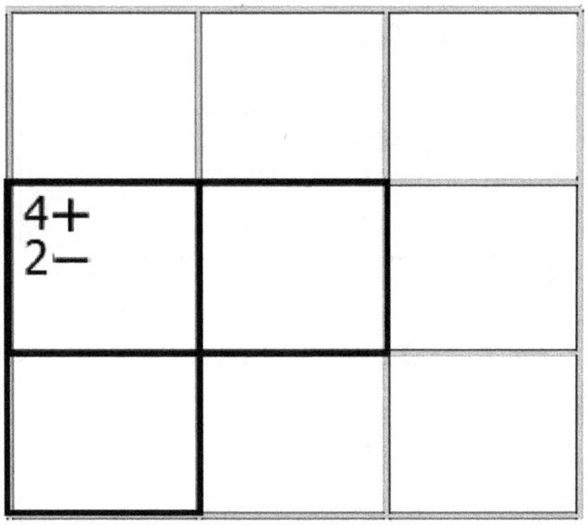 4+ 2−	The left CalcuDoku diagram causes confusion because we do not know which box is for 4 + and which box is for 1−. The Venn diagram concept can be easily demonstrated in the *Frankho ChessDoku* but causes confusion in the CalcuDoku.
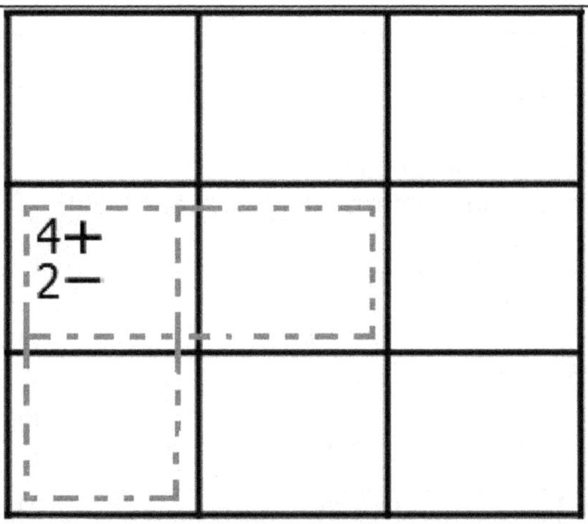 4+ 2−	The left CalcuDoku diagram uses the dotted box, but again it still causes confusion as stated above.

Student's name: ________________ Assignment date: ____________________

Commutative law

The traditional way of calculating is in the direction of left to right or top to down. Still, this rule does not apply to CalcuDoku because, as shown below 2 – can be expressed as 3, 1 or 1, 3, and it appears to students that the subtraction can be done by exchanging the two numbers, and this is in violation of the commutative law. It would have no problem for *Frankho ChessDoku* to handle the subtraction and division operators because the calculation direction is clearly defined by using chess moves.

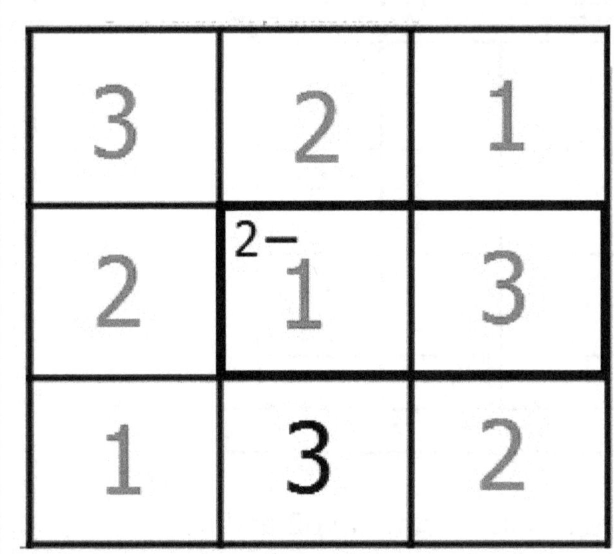

3	2	1
2	2– 1	3
1	3	2

The answer could be operated from left to right for the subtraction operator, but sometimes, it could also be from right to left.

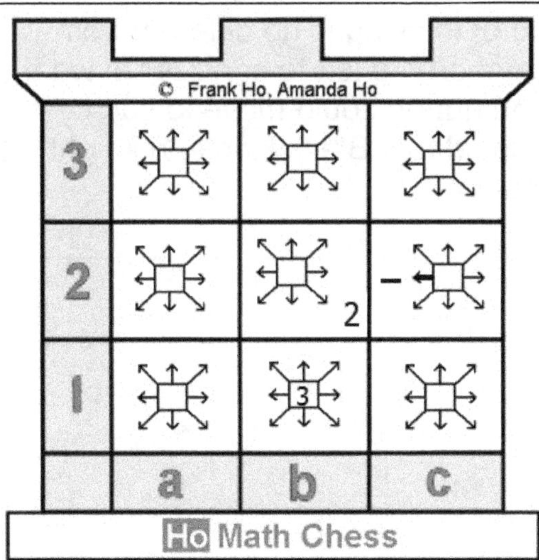

The above CalcuDoku requires the student to think about how 3 and 1 are to be arranged to present extra challenges for students. Still, the confusion could also occur when the mixed operators (+, –, ×, ÷) are presented together with no operating directions are given.

The left *Frankho ChessDoku* presents no operation confusion and does not confuse commutative law for children.

Student's name: _______________ Assignment date: _________________

Chess strategy and *Frankho ChessDoku* strategy

Often a chess player would analyze the chess moves and see where each chess piece intersects with each other, then decide to take the next move. This kind of thinking is also reflected in the strategy on how to solve *Frankho ChessDoku,* and the following example demonstrated the transferred knowledge between chess and *Frankho ChessDoku*.

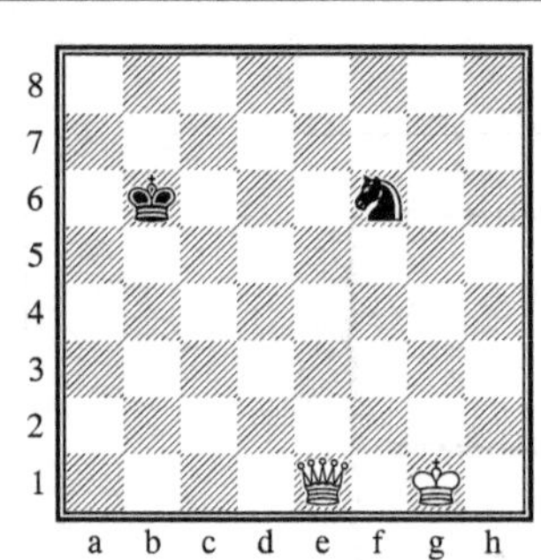

Find a Black move to fork.
Qe1 moves to f2 to fork black king and knight.

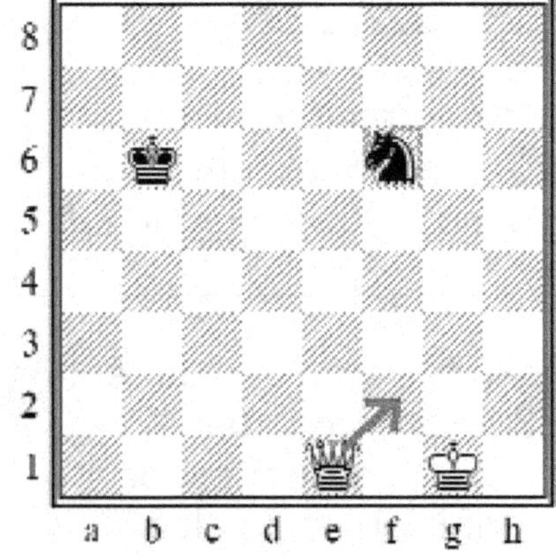

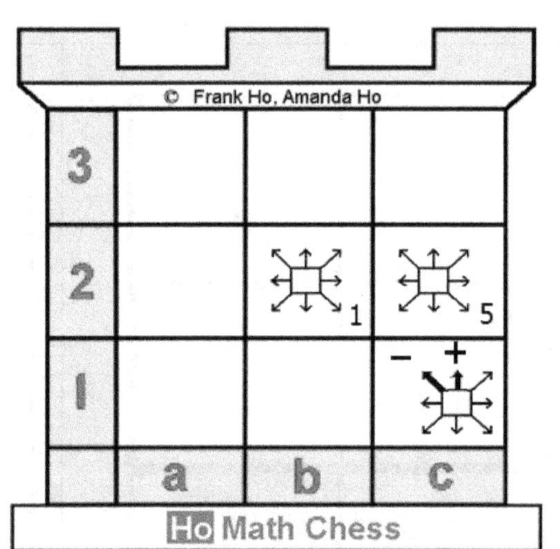

The above c1 intersects with b2 and c2 simultaneously, so in other words, c1 is a square where bishop and rook intersect in chess. This kind of thinking is no different from the chess diagram on the left to consider at what square where the queen could move to such that the queen could fork Black king and knight at the same time.

Student's name: ________________ Assignment date: ____________________

Triangular solving strategy for 3 by 3 grid

The simplest 3 by 3 case of *Frankho ChessDoku* can be created using only one number and one math operator. All other math operations are really redundant.

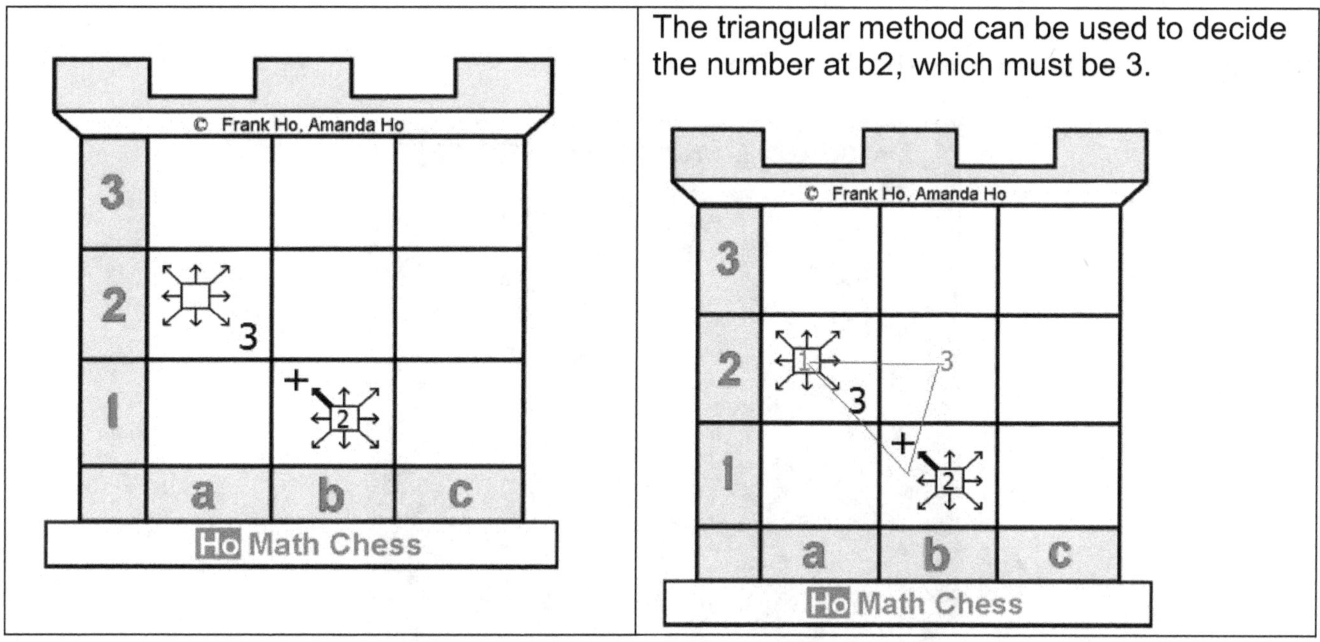

CalcuDoku cannot duplicate the above diagonal operation, but a vertical and horizontal operation can be made.

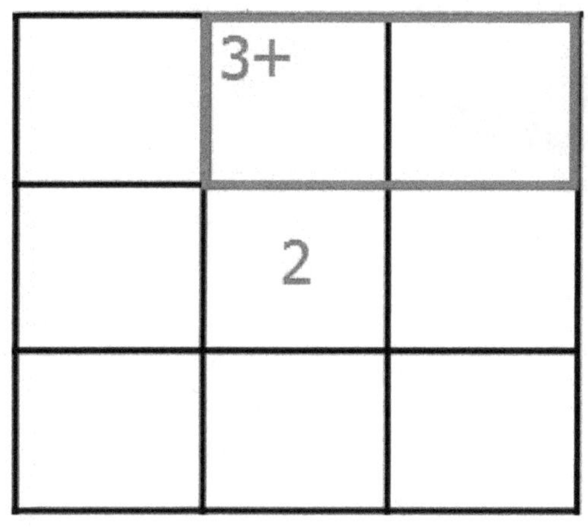

Frankho ChessDoku

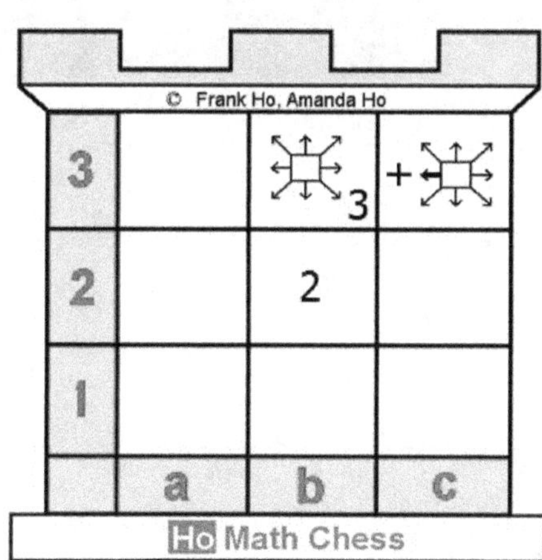

Student's name: ________________ Assignment date: ____________________

References

(1) http://susanpolgar.blogspot.ca/2008/10/ho-math-and-chess.html
(2) http://www.mathandchess.com/releases/release/1441781/17465.htm
(3) http://en.wikipedia.org/wiki/Tetsuya_Miyamoto

Students can frame their Frankho ChessDoku as follows.

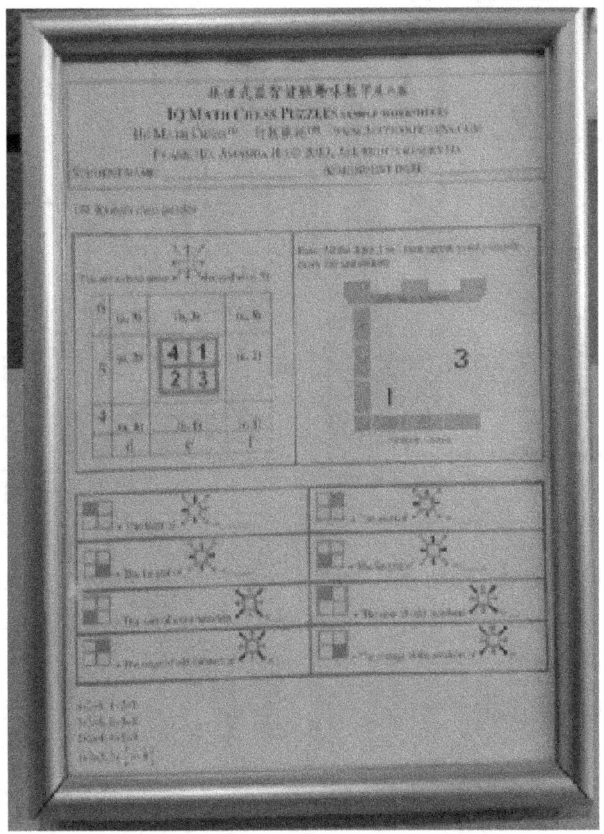

Student's name: ________________ Assignment date: ____________________

An Innovative and Fun Way of Teaching Math

I have observed that one of the main reasons children hate math is they have not mastered the foundation of math; surprisingly, the root of problems can be traced to the basics they learned in the elementary grades, such as addition, subtraction, multiplication or division, when elementary students are asked why do we have math. Many of these children can relate math to their daily lives, such as shopping, cooking, measuring, driving, time, shapes, etc. Most of the time, they think math is important because it has something to do with counting and numbers.

It is terrific these youngsters realize that math has relations to our daily life but does the traditional way of doing math worksheets such as 2 + 3 reflect the environment today our youth lives? It certainly does not. Math is not just about counting numbers. Besides teaching the basics of addition, subtraction, multiplication, or division, math is supposed to teach our next generation how to solve problems and be creative. Part of the problems why some children hate math worksheets is simply because these traditional worksheets do not represent the world they are living in now.

Many "things" or "toys" our generation play already start to reflect what the society would be like in the future, things such as internet searching and browsing, image viewing, cell phones, roaming, instant messages, downloading and uploading information, compressed files, file formats etc. all of these will become part of their life and necessary living skills. How do math worksheets reflect the way children are living now or will be in the future? Most school math textbooks still use the same worksheets to teach children basic math that is basically number crunches in the straight top to down or left to right fashion. There is very little convergence of numbers, images, patterns, comparisons, searching, matching, sorting, classifying between one number operated with another number. This "old" way of manipulating numbers does not represent what our children are doing in today's society. This may have explained why most young children still only think of math as counting numbers, but not related to how they process a variety of information in their daily living.

Ho Math Chess simply uses international chess as a tool to converge the information of images, patterns, sorting, comparison, matching, tables, etc., all together with numbers to better reflect what children are facing or already doing today. We have achieved this learning in a fun way.

Is there such a thing that math worksheets can be created to be fun-oriented that children can play with numbers and be rewarded with satisfying and great fun? How can the dreaded drill style of math worksheets be improved such that they encourage whole-brain learning? How can math basics computations such as addition, subtraction, multiplication or division be incorporated with the fun gamed-based approach? Is it possible that children can improve their memory and problem-solving skill in a game-based learning environment?

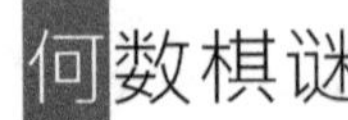

Student's name: _______________ Assignment date: __________________

Ho Math Chess has created math and chess integrated worksheets to give positive answers to all the above questions. So how does Ho Math Chess put math, fun, thinking, creativity, and memory improvement all together?

Chess is a strategy game, and it invigorates the thinking ability and has been around for thousands of years. Many organizations and schools have set up chess lessons, and often math and chess are taught under the same roof but with no real integrated math and chess curriculum. Math and chess have traditionally been taught separately because there is no existing connection mechanism between math and chess teaching. However, since Frank Ho, founder of Ho Math Chess, invented the copyright applied Symbolic Chess Language (SCL). Ho Math Chess has successfully created the world's first math and chess integrated workbook using its proprietary intellectual product (SCL).

The future belongs to a generation that understands how to process information, and the information might include digits, bytes, numbers, graphics, images, languages, symbols, equations etc. How can these different natures of information processing be taught to kindergartners or primary students when learning arithmetic? It is not an easy task; that is why there are so many different types of standalone worksheets: logic, patterns, mazes, or crosswords. These worksheets are created without interrelations with each other. This kind of isolated information processing is no longer reflecting the young generation's real-world today or living in the future.

The computing world children are facing today is much like a rich tapestry, where diversified fabrics and colours are integrated. Children today are absorbing not just numbers but an array of information like images, sound, music, symbols, spatial information, or even abstract ideas all bundled together and delivered through many types of media. Children today are not happy just working on pure number drills without any other stimulus or motivator.

Realizing the importance of having fun while learning, Ho Math Chess has been embarked on a crucial teaching philosophy that is to integrate chess into math worksheets so that children can learn math while having fun.

Ho Math Chess created a special synergetic effect by integrating arithmetic basics, chess, mazes, and information processing all on one worksheet. This is accomplished through its proprietary technologies such as SCL, Frankho ChessDoku, Frankho ChessMaze, and an innovative Ho Math Chess Teaching Set.

Student's name: ________________ Assignment date: ___________________

With the new invention of Ho Math Chess worksheets, a child is acting as a data warehouse manager and sorts data through a variety of tools, namely chess, symbols, spatial relation, logic, comparison, tables, patterns, mazes, computing etc. by networking all kinds of information together. Only when children have successfully followed through instructions (SCL) and, as a result, created a question themselves can a solution be found at last.

In Ho Math Chess worksheets, the questions are not written out for children but must be mined (after children observing how data is moving) through the data warehouse (mazes). Answers must be computed after conducting a series of logical thinking processes, including spatial relations, sorting, comparing, matching, classifying, tabling look-ups etc.

Ho Math Chess trains children their basic computing ability and trains them to be astute data warehouse managers or excellent data miners by developing their problem-solving ability and critical thinking skills.

Ho Math Chess provides education and also entertainment value to get young children involved in the future world they will be facing.

At Ho Math Chess, math learning is fun.

Student's name: ________________ Assignment date: ____________________

Three reasons why students can't do math well

I was curious to determine why a child who is already grading five but could not do basic multiplication well despite explaining concepts. I was equally at a loss to try to find out why a grade 10 student who could only "see" how a problem can be done after I explained it three times or write the solutions step by step and also allowed him to stop me at every step whenever it was necessary. In this case, I know the grade-10 student never 100% understood the math instruction in his day school because he could not stop his teacher's instructions just for him on every computation step. For example, When factoring (x square – y square), the student just could not see how this problem can be done when the problem changes to (x + 1) square – (y+1) square.

After I did my observations and thinking, I found out these problems have lots to do with their math training while they were young other than their ability deficiency in basic math. They lack training in the following areas:

1. Pattern: They have some difficulty in seeing patterns.
2. Visualization: They have difficulty to see the relationship between columns or rows or table to table or from numbers to graphics and vice versa.
3. Abstract: They seem to have difficulty grasping the abstract idea. Any problem requires students to convert using abstract ideas. Then they seem to run into problems.

To take care of these problems mentioned above, Ho Math Chess has created a series of workbooks to train children in patterns, symbols, and relationships using visualization.

We do it to use our patent applied Geometry Chess Symbols and have created a series of one-of-its-kind workbooks such as IQ puzzles, Frankho chess mazes, IQ mathematical chess puzzles, word problems, and math contests to increase their brainpower in number intelligence enormously.

More information on how to improve children's math ability can be found at www.mathandchess.com.

Student's name: ________________ Assignment date: ___________________

How to raise math competition ability quickly

A student's math competition ability takes time to develop. The math competition ability is raised gradually through continually working on math problems, but is there any way that a student's ability (say the students are already at grade 6 or grade 7) could be raised more quickly than normally could have done when the student had not had the time to work on some math competition problems at a younger age? In other words, how do we train a late bloomer to prepare for math contests?

Suppose students could learn the following math concepts as earlier as possible after building the 4 basic operations including whole numbers and decimals. In that case, they will possibly do much better in math contests. This idea can easily be transferred to prepare students interested in preparing for any math entrance exams such as SSAT or enrichment math programs.

1. Know how to work backwards, especially in fractions.
2. know how to use the concept of 1 or 100%
3. Can comfortably convert the three types of data among decimals, fractions, and percents without any problems.
4. Know the concept of quantity divided by its corresponding fractional number and also the concept of fractional numbers.
5. Learn fractions right after learning the four basic operations. The reason is that converting an improper fraction to a mixed fraction involves the concepts of division, multiplication, and addition all in one operation. Also, it teaches the concepts of why dividend is found by using the divisor × quotient and then adds the remainder. An equation concept is learned at this stage also.

There are a few computation techniques that can be sued to make computations much easier, and they are listed as follows:

Student's name: ________________ Assignment date: ____________________

1. If the answer for the fraction is $\frac{7}{1}$ then it should be written as 7. However, when doing the following computation, it will be a different story.

   $\frac{7}{3} \div 7 = \frac{7}{3} \div \frac{7}{1} = \frac{7}{3} \times \frac{1}{7} = \frac{1}{3}$

   Why use "Invert and Multiply" in fraction division?

   $\frac{7}{3} \div 7 = \frac{\frac{7}{3}}{7} = \frac{\frac{7}{3}\times 1}{7} = \frac{7}{3} \times \frac{1}{7}$

   Use the Unitary method. We know there are 7 of $\frac{1}{3}$, so when it is divided by 7, then the result is just $\frac{1}{3}$. From this example, we know that a fraction can be a "division" such as $\frac{7}{3}$ operation and also a "multiplication" operation such as $7 \times \frac{1}{3}$.

2. When encountering fraction division, multiplication is performed. When encountering a multiplication, a division is performed; for example, the following $24 \div 12$ is performed first before multiplication.

   $24\times \frac{5}{12} = 2 \times 5 = 10$

3. A fraction involves left to right order of operation, top to down operation, and diagonal operation. The following example demonstrates three operations.

   $\frac{24}{12} \times \frac{7}{5} \times \frac{15}{21} = ?$

   The reducing operation involves top/down and diagonal, and the final result is obtained by multiplying numbers across from left to right.

4. Because the fraction operation is so powerful, the word problems of % or ratio are transferred to fraction, and the fraction operations are performed to solve the problem.

Student's name: ________________ Assignment date: ____________________

5. We must learn how to translate English sentences into math sentences.
   Example 1 For every three girls, there are two boys.
   It could be thought of as a repeating pattern problem such as the following:
   BBBGGBBBGGBBBGG......
   It could also mean the ratio of boys to girls is 3 to 2.
   It could also mean there are 3 boys for every 5 children and there are 2 girls for every 5 children.
   Example 2 Find a number less than 100 and is divisible by 5, 6, 7, and 8.
   This problem just means to find LCM of 5, 6, 7, and 8.
6. Take one third means divide by 3.
   For example, One-third of a number adds 30 are 66. What is the number?
   $x \div 3 + 60 = 66$
   $66 - 60 = 6$
   $6 \times 3 = 18$
7. Often a math operation can be "seen" clearly when converting to fraction operation.
   $9 \div 5 \times 5$
   $= \frac{9}{5} \times 5$
   $= 9$
8. Take care of negative sign and reduce as early as possible to avoid troubles.
   Example
   Compute $\frac{-4}{2} \times \frac{22}{-11} \times \frac{121}{-11} \times (-\frac{169}{13})$

Student's name: ________________ Assignment date: ____________________

Problems students could face when calculating with fingers

There are many ways to train children to calculate. Some teach abacus, and some use the unique finger method. Ho Math Chess worksheets' design encourages students to do calculations without using their fingers to do the counting forward or backwards. None of our students and students participating in math competitions use their fingers to do necessary calculations. Their mental math ability is strong than those who use fingers to count. The main reason is that, without using fingers, students will develop a strong number sense and see steps ahead when working on math problems.

The following gives a list of potential problems students might encounter if they continue to use their fingers when going to higher grades:

- Difficult and slow to find answers related to work backwards, such as subtractions or finding quotients. For example, it takes finger calculating students much longer than peers to find 12 - 9? When doing simple divisions such as 24 ÷ 2, 36 ÷ 3, and finger calculating, students cannot do them mentally without using a pencil and paper.
- Have a hard time to find the product of primes of a number; slow to find an answer and have a hard time to see if a number is divisible by a certain number using divisibility rules; challenging to find GCF or LCM and later LCD; difficult and slow to reduce a fraction; have a hard time to do the conversions between fractions, decimals, and % etc.
- Unable to see what is the trick to add certain type of questions such as 99 + 7 + 8 + 4 + 6 + 1 + 3 + 2 etc. Students who do not use their fingers can scan this problem, and they know the answer almost immediately.
- Because of the slow speed of doing basic calculations and the lack of mental math ability, the student has less chance to get into the enrichment math program due to competitiveness. The student has less chance to do reasonably well in any math contests, which stress calculations speed and the calculator is not allowed. Students will do poorly when oral math problems are presented in any math contests.
- Students will continue to experience difficulties when going to high schools with those problems requiring mental math quickly to recall four basic computation facts. Problems include finding the square roots, adding like terms; factoring trinomials using cross-multiplication method; 30-60-90 special degree, graphing $2x + 3y = 6$ etc.

When we talk about calculation without using fingers, we are not saying the only benefit is so that students can recall arithmetic facts quickly. **From the above observations, we can see that mental math or calculation without using fingers is not limited to the scope of being a human calculator of doing some calculations in lighting speed.**

Student's name: ________________ Assignment date: ____________________

To acquire the mental math skills, so it truly benefits students, the students must learn the skill of reversing calculating like how to get factors of a number, how to quickly find prime factors of a number, how to find if a number has a perfect number as a factor, how to not only do the calculation in a linear fashion but also know how to do calculations in cross or vertical way quickly and be able to see patterns. Mental math also helps students do trinomial factoring when they are in grade 9 or 10.

When in high school, students with math deficiencies often find it is difficult for them to go back to work on addition, subtraction, subtraction, or multiplication again since it is embarrassing for them to work on grade 3 math while they already grade 10.

With the above in mind, we shall create an environment where students are encouraged to think more even when they are doing calculations. We also shall encourage students to do calculations without using their fingers, so the calculations are totally done in their brains without relying on any manipulative. Calculating without using fingers can help students develop strong mental math abilities. As a consequence, this helps students do well later in their SSAT, SAT and also increases the chance of getting into private schools if they want to. The chance of getting into an enrichment math program in high schools is also higher.

Student's name: ________________ Assignment date: ____________________

My experience of teaching children with math disability or dyscalculia

I taught a few young children who had math disabilities or dyscalculia, and they were around 4 or 5 years old. They had experienced tremendous difficulties in learning math and it was a challenge for me also to teach them. Below is one case.

He had problems in writing numbers and counting numbers, so right from the start, he already experienced difficulties before getting into actual computing, such as simple additions or subtractions.

I searched many workbooks on the market and found none of them was suitable for him to work with, and I was forced to create a workbook which I wrote to take care of his problems.

His problems are as follows;

1. Writing

He had a problem with writing the correct numbers. He could not distinguish curve and straight lines when writing numbers or wrote a digit 3 with an extra curve. He needs more worksheets in practicing his writing because just a few pages of tracing numbers are not enough for him. He is very confused about 2 and 3 because he was not sure if 2 has one-half circle or two halves of circles.

2. Reciting and Counting

He could not recite numbers fluently forward or backward and could only recite from number 1. If you ask him to count not from 1 then he may have problems. He could, on the one hand, to count from 1 to 4 objects, but then he would write the answer as 3 objects without realizing he wrote wrong even when given the opportunity to correct and after the condition was explained to him.

His understanding of why his answer was wrong is not high. He could not count backward from 10 to 1 correctly by correctly saying the correct words. He also had speech problems.

Even he can do 1 + 1 = 2 but, all a sudden, a few minutes later, he would give the answer 1 + 1 = 1.
He seems to do problems by 'remembering', not by reasoning. When using cubes to show him 1 +1 =2, and he understood, but if changing to a new blank sheet and wrote 1 + 1 = ? He got confused. And would give answer 1 + 1 = 1 because he has been given answers of 1 and 3 for other problems, so he just went ahead to give answers as 1 or 3.So by observing him, I feel that it is very important we train him in reasoning and try to get him to understand the reason behind of all his answers.

Student's name: ________________ Assignment date: ____________________

3. Logic and pattern

He had difficulties in identifying patterns, and the pattern almost does not exist to him. Any number or objects which require some logic will be very difficult for him. He could not understand even after he was taught, so his understanding of logic is not high. From this point, I feel that he needs to be trained in logic and reasoning.

4. Retention and review

It is not really his 'right answer' or 'wrong answer' that worries me, but the ability that he solves the problems worries me, and it leads me to believe that he seems to have a math disability. He can do well after repetitive instructions and but just a few days later, he would suddenly act in a way that he seemed to have never learned before. He does not seem to have any retention. Also, once he thinks something is right, then despite my teaching, he is unable to change himself. For example, after I pointed out to him that 5 is a correct way of writing and the following writing of 5 that he wrote that way is incorrect (as follows) and I asked him to write and traced 25 times, but at end, he still wrote his way.

Student's name: ________________ Assignment date: ____________________

A secret to helping your child succeed in math

We can use many "secrets" to help our children make math a successful subject, but what if I were allowed to use only one secret? What would be the only secret that I could use to make math a successful subject for my children? The only secret I would use would be to train my children's thinking skills, not hand computation ability. Why do I say that? The reason is too many times. I see my own tutoring children who cannot perform simple computation in their brain without using the "pencil and paper" method. It is not only slow but also error-prone, and they did not know it. Many of these children have failed to use every opportunity they could to train their brain and exercise their brain. Because of this, I feel it is very important that we stress mental math is important when kids are young but continue to stress. It is equally important when they are older, going to high school. The trouble is many tutors can "see" it is important to get the fluency of 4 basic operations (+ -, x, divide) but failed to see how children should be taught about mental math when they are older.

I give a few examples below to illustrate some of the failing jobs that math tutors have not stressed the importance of mental brain training when kids are older.

1. When testing roots of a function such as a sub one into $x^3 - 2x^2 + 3x + 4$, it is very easy to calculate f (x) by sub 1 into f (x) and get answers mentally. Many students are doing it by hand. X $(1)^3 - 2(1)^2 + 3(1) + 4$. It is very "easy" for students to get the power of 1, so there is no need to even write the computation expression to try to figure what is (1) ^3 or (1) ^2 etc. Encourage students to do this kind of job mentally whenever they can to train their brain.
2. $2x = 4$. Do we have to ask students to write it $2x/2 = 4/2$ by hand to get an answer x=2?, Can't they "see" that both sides dividing by 2 so they will get x=2 and do this in their brain? Again, we shall encourage students to use their brains to do this - it is faster, and they get a chance to exercise their brains.
3. When calculating the root conditions of using discriminant, it is unnecessary to actually get its value other than > o, < o, or = o. Many students are not able to "see" or "judge" its result as positive, negative, or = 0 without actually calculating its final result. Again this is the sign that their brain has not been trained to do it mentally.

We must encourage students to compute mentally whenever they can, not encourage them to do computation always by hand and paper.

Student's name: ________________ Assignment date: ___________________

Ho Math Chess creates revolutionary math workbooks

There are three important objectives that must be achieved to run a successful math specialty learning centre. The most important one is the student's day school math mark must show improvement. The other two are parents must be able to see the teaching value, and the children must like the program. With the approval of the trademark of Ho Math Chess Learning Centre's invention of Geometry Chess Symbol, we are able to create some disruptive, revolutionary math workbooks which are out of this world, and we are also very pleased to announce that we are able to achieve these three important objectives as mentioned above with the creation of our unique products. We are able to reach an important milestone because of the following reasons.

1. The founder of Ho Math Chess, Frank Ho, discovered the secret code which links math and chess by inventing the Geometry Chess Symbols and also trademarked the symbols. As a result, many products such as Ho Math Chess Teaching set, Frankho ChessDoku and Frankho ChessMaze were created.

2. Ho Math Chess is the first learning centre in the world to create math, chess, and puzzles integrated workbooks.

3. Ho Math Chess also has developed the workbooks to boost children's brain power using math, chess, and puzzles integrated technology.

4. Ho Math Chess has created enriched material for more advanced students.

5. Ho Math Chess invented the Ho Math Chess teaching set specially designed for young children; thus, kids as young as 4-year old can learn chess much faster than they are using the traditional 3-D chess set.

6. Ho Math Chess created special math and chess integrated workbooks for kindergarteners.

Ho Math Chess materials make children like to learn math when compared to the use of traditional math worksheets. At Ho Math Chess, not only do children learn math; they also learn chess and work on puzzles to improve their brain power at the same time.

Student's name: ________________ Assignment date: ____________________

How to Play Blind Chess or Half Chess

Ho Math Chess Teaching Set is specially designed for young children to learn chess. Since Ho Math Chess Teaching pieces have flat surfaces and uniform outlook in square size, the pieces cannot be identified and are indistinguishable from each other when they are turned face down.

This special feature allows children to play a special game called Blind Chess or Half Chess (or called Banqi in Chinese Chess). The rules to play Blind Chess are very similar to Chinese Blind Chess. Blind Chess is very easy and fun to play.

Board

Blind Chess is played by two-player on half (4 by 8 square board) of the normal chessboard.

Game Rules

The 32 pieces are shuffled, and then each of them is randomly placed face-down on each square. The first player turns over a piece, and the colour of the first piece uncovered will be the side of the first player.

Moving a Piece

There are 3 kinds of moves. A player may turn a piece face-up, move a piece, or capture an opponent's piece. A player may only move face-up pieces of his or her own colour. Unlike normal chess moving rules, there is one rule to move pieces in Blind Chess: a piece moves only one square up, down, left, or right. Namely, all pieces move like a rook. A face-up piece may only capture a square occupied by an opponent's face-up piece.

Student's name: ________________ Assignment date: ___________________

Capturing an Opponent's Piece

All pieces (Black or White) are ranked according to the following hierarchy, and the capturing rule is strictly according to the defined hierarchy.

King has the highest rank and can capture the opponent's all pieces other than a pawn.
Queen can capture the opponent's all pieces other than a king.
Rook can capture the opponent's all pieces other than king or queen.
Bishop can capture the opponent's all pieces other than the opponent's king, queen, or pawn.
Knight can only capture the opponent's pawn.
Pawn has the lowest rank but can capture the opponent's king.

How the game ends

The game ends when a player cannot make a move or until all pieces are captured. If the game is forced into an endless cycle of moves, then it is a draw.

Student's name: ________________ Assignment date: ____________________

How to raise math smart kids

Over 15 years of math teaching, I have observed that parents' involvement in their kids' involvement has a great influence on their progress, especially when they are young. Parents can help kids to learn math as young as they can start to hold a pencil and can start to write. So I am talking about kids as young as 3 to 4 years old.

Why are these kids so smart in math when they are as young as 4 or 5 years old? I have noticed that their parent's attitude is very different; let me outline my observations as follows:

1. They are extremely patient; they sit beside their child and study along with their kid with 100% devotion. They do nothing but study together with their kid.
2. They do not just want their child to do the only computation. They want their child to learn also math problem solving and especially the materials which can train the brain.
3. They come to my class in person every time dropping a child and come into the office every time picking up the child and always greet me and chat with me a few words.

Student's name: ________________ Assignment date: ____________________

Case Studies on How Math Chess Affects Children's IQ

In my over ten years of teaching math and chess, I have had the precious opportunities to teach grades from kindergarten to grade 12, and as a consequence, I also have had many occasions to observe some interesting cases on how some of my own students learned. After I analyzed their patterns of learning, I have used their learning experiences to modify Ho Math Chess worksheets.

Case 1 – 4 years old boy

My feeling is that we seem to use different parts of the brain to do math and chess, respectively. This observation is based on the following case. If this is proved to be true, then it definitely is beneficial for children to learn both math and chess since different parts of their brain power will get "strengthened".

This 4-year boy really surprised me when he came to me since he already knew how to add or subtract 1 digit quite comfortably without any noticeable delay in coming up with answers and could also do 2-digit addition or subtraction, although my experience with him was not sufficient to draw any conclusions. He also knew the number skip pattern and could do multiplication as well. His mom wanted me to teach him chess, so I started with chess lessons, but it caught me by surprise was his progress was slow when compared to the progress of his math ability. Why is there such a disparity? Every time when I played chess with him, he needed to be reminded about how each chess piece should move even after he had shown that he already knew the moves of each piece. He did not seem to understand that to play chess well. He needs to look at the entire chessboard, not just one move of one chess piece. He did not seem to care or understand the "what if" effect.

This smart boy intrigues me to think why he could do calculations so well and yet is slow in absorbing chess knowledge, so I started an experiment by giving him a word problem for he could read already at four years old. I would explain to him if he did not understand the meaning or could not read some words. I found out that if he was capable of doing something, then he would do it quickly, but if something he could not do and I tried to explain to him and he still could not do, then he would simply "shut off" his brain and would not do it, so in this case, I was not able to observe how he could make progress under my guidance.

Perhaps, the function of playing chess is being handled by one part of the brain, and the functions of a mathematical function are mainly being handled by the other part of the brain, so in this boy's case, he could do math computation very well, but not chess.

Student's name: _______________ Assignment date: _________________

Case 2 – Grade 2 Girl and a grade 3 boy

From these two students, I learned that children could "shut off" their brains and refuse to learn. I taught this very friendly girl when he was in grade 3, and now she is in grade 8, and I am still tutoring her. I tutored the boy when he was also at grade 3 level, and then I later met him again at grade 8. I asked both of them the same question why they did not seem to be able to master the times table when they were young, and their answers astonished me since both of them said to me they just did not want to "memorize" it.   I have a more interesting story to tell on how they overcame the problems themselves and how this little girl influenced me in creating my workbook.

In the girl case, I seem to have hit the roadblock since it does not matter how I taught her by using flashcards, explaining the multiplication concept, using pencils as manipulative she just could not do times table, and finally, I realized that she understood the concept, but she could not come up with the results with no delay, or she would just come up with wrong answers. At no time I would "force" her to memorize the times table other than to remind her that she might as well spend some time reciting and memorize them. It never happened to her. I reminded myself that this is a great chance for me to produce a custom-made multiplication table for her to see if she could actually use my multiplication table to "naturally" "memorize" all multiplication facts. I asked her what kinds of multiplication questions she liked and what kinds of multiplication questions she did not like. My multiplication workbook went from initially with 100 pages to finally over 300 pages because of working with her. Clearly, I learned that she did not like simple drill type worksheets because they are boring. I was "forced" to come up with a different variety of formats to suit her taste. Finally, she could do multiplication but with some delay in coming up with answers. When getting to higher grades, her computation ability seemed to have a leap of progress, I asked her what had happened she told me that she just decided to "memorize" them and that is it. The same happened to the boy, I asked him why he could not seem to get the times table, he told me that he just did not want to memorize it when he was young, and later, he decided to memorize them.

So have we ever paid attention to the fact that some children could not learn times table well was because they did not want to "remember' them? Educators did not seem to do this kind of "follow-up" study in education to find out what was the problem when some children could not learn when they were young.

Case 3 – Ho Math Chess worksheets vs. traditional worksheets

From my own teaching experience, I have found that for some children, the drill is boring and causes stress to children. On the other hand, drill gives the fluency children need to do math by hand, and the skill children need to do word problems with second nature and be able to "see" the direction of getting answers.

Student's name: ________________ Assignment date: ___________________

What's wrong with drill rests with the tool of drilling that is the worksheet. It is not interesting, boring, and monotonic. It makes children feel like working on an assembly line to get all those answers. I personally gave Ho Math Chess worksheets to children and traditional worksheets to children, and clearly, children liked Ho Math Chess worksheets more than the traditional sheets. I also asked children why they liked Ho Math Chess worksheets, they gave me the reason that they liked to work through a few turns and jumps to come up with the answer, and they felt it is more interesting that way. When I heard this kind of comment, it boosted my confidence that I have come up with an innovative product which revolutionaries the traditional worksheets. My invention of Geometry Chess Language is a disruptive technology and an innovative idea that basically changed forever how children should be drilled and how to make math worksheets more interesting. I was almost holding back my tears when my students were telling me why they liked Ho Math Chess worksheets.

Case 4 – Some children do not mind, but some do not like to be drilled.

This is a true story that some moms told me that they could not send their children to other learning centers anymore since their children are in tears when asked to go again – they hated the drill. On the other hand, I see, although rarely, some children do the drill worksheets until grade 12. Why is there such a big difference? I started to feel that our brain seems to function differently depending on the individual and for some reason. Those children hated drill is because some part of their brain is telling them to "reject". We can do an experiment by giving Ho Math Chess worksheets to the students as a group and the traditional worksheets to another group of students and seeing how their brains have reacted by scanning images of their brains. This will remain a mystery until someone has done some research in this area.

Student's name: ________________ Assignment date: ____________________

Characteristics of Math A Students

This article attempts to see if there are some characteristics differences between 2 groups of students: those students who consistently getting A in math and those students who could not consistently be getting A in math. The author believes understanding the characteristics difference between them is an effective way to raise each individual student's math performance.

Let's say there are 20 students in a regular math class, and there is most likely a few who consistently reach over 90% performance in math tests, so why students studied from the same teacher and were asked to do the same amount of homework and listened the same lecture ended up with different test results? There are a lot of reasons, but can the non-A students learn something from the A students and see why the students consistently get A? The trouble is most students cannot "see" how A-students study differently in comparison to non-A students. I personally tutor math from kindergarten to grade 12 students, so I have had the opportunities to observe how A student perform differently from non-A students. The difference is apparent enough to conclude that some students cannot get A is because they themselves have not got the same characteristics of learning habits that most A-students have possessed to achieve the A status.

Some of my observations are as follows:

If the students already have the knack for math and they also study very hard, then certainly it is easier for those students to get A consistently. On the other hand, students who have average intelligence but with good study habits and study hard can also achieve A, but extra work is required. Some non-A students just do not work hard enough and are not willing to work on extra practices. It is difficult to get an A for those students who somehow just cannot get it despite tremendous efforts of teaching using models and manipulative. With that in mind, it seems to be unfair to use the statement that everyone can learn math well and get A.

Math A students care and do not forget what they have been taught. Recently, I have taught a grade-5 student on prime numbers by asking him to write all primes from 1 to 100 (after my explanation of prime number.), and he complained. Later I gave him 20 or so some prime number questions to practice and also reviewed his mistakes. Despite the efforts I prepared for him, He could not get perfect on the following two questions given by his schoolteacher:

List all the prime numbers between 36 and 46. List all the composite numbers between 36 and 46.

I was disappointed that he missed one number in each of these questions. I was wondering why he could not complete it with perfection since these two questions are not difficult. My feeling is my efforts to study with him on primes shall enable him to get the above questions totally correct, but he didn't, so what was the problem?

Student's name: ________________ Assignment date: ________________

I noticed that this boy could not put in his energy 100% whenever I was tutoring him. He consistently had to get up his chair for some reasons and also asked me to give him time to play chess; waiting at his seat without doing anything when not told to do specific work (i.e., no proactive attitude); complained when given work to do; chatting with neighbours whenever he could. So perhaps he was not even paying attention when he was asked to write all primes from 1 to 100 at the time I was teaching him, or he was not paying attention when he was taking his test? All these are related to him. Personally, it really has nothing to do with the math curriculum or teaching method since we are only looking at one small problem and find out why he could not get it 100%.

I personally feel that some students could not get A is because they have not got the characteristics of A students' good study habits.

Non-A students do not bother to retain anything related to concept teaching. If students do not bother to remember any concepts, then how can they do well in math? I told students that learning math is not like massage work. You cannot just sit there and let the tutor does the work. Many non-A students treat learning math like it is only the tutor's job to do the work and perhaps understand at the moment when the tutor was explaining, but next week, when they were asked again to do the same practice, they acted as they have never been taught.

Non- A students left all notes on the table after teaching and did not care if they can review it at home or later.

Math-A students will take the proactive attitude to tell me what they did at school and "guide" me to the area, which they do not understand, but non-A math students do not care.

Math-A students ask me something they do not understand in the class, but non-A students have no questions to ask me.

Academic performance has lots to do with a student's personal living habits too. One cannot be lazy and must be willing to and prepare to work harder. Math-A students must be willing to use their brains and cannot be lazy; one small example is to see if students are willing to use their brains to arrange food on the refrigerator's shelves in a very organized way. Open the refrigerator and see how some of the students place the food on the refrigerator's shelves. Have they used their basic math knowledge to maximize the shelves space?

Non-A students do not consistently finish their homework completely and do not feel shame at all is another reason that some students do not do well, and even worse is some parents are making reasons to forgive their children.

Student's name: ________________ Assignment date: ____________________

The common reason given by young students for not doing some math is they complained to me that this is not what they do at school. Not willing to take on any challenges is another problem facing this kind of student, and often they will use the same reason to complain to their own parents.

Have a goal, care about what you learned in class, help tutors by pointing out what you do not know, do all homework. Be organized on math notes and collect all notes left by the tutor and be willing to review them at a later date. These attitudes seem to be all just common sense to be "good" math-A students but lack in some of these attitudes is exactly the reasons why some non-A students are not getting A.

Student's name: ________________ Assignment date: ____________________

Chess for Math Curriculum

Background

A myriad of education research papers have concluded that chess benefits children in many areas and one of them is that it increases math scores (1, 2). This conclusion begs an inevitable question: how does chess seem to improve children's math scores? The most obvious reason one could probably think of is that there is a connection between chess and math, but how are math and chess connected? In addition, how are these connections actually linked to the school curriculum?

John Buky has outlined the chess curriculum aligned with the National Mathematics Standards (3). This article sets to find out the links between chess knowledge and school math curriculum learning outcomes. The other purpose of this article is to further evaluate how mathematical chess puzzles could reinforce math concepts learned and thus increases the potential of improving math ability. The process of play a good chess game is similar to the way of how to conduct critical thinking and problem solving, and this connection is well known, so this article will not research the connection between chess and critical thinking skills.

A further exploratory question would be that if chess provides benefits in improving math ability, then would it even more beneficial if children could actually work on some mathematical chess puzzles?

Method

A comparison is made between Canada BC Ministry of Education mathematics Grades K to grade 7 Learning Outcomes, and chess knowledge learned. The comparative results would show what chess knowledge learned would match the learning outcomes of the BC Ministry of Education mathematics curriculum. This close relation demonstrated should provide strong evidence to the reason why chess improves math scores.

Comparison Summary and Observations

See Appendix A for details on the comparison. Below are a summary and observations.

Grades K to 4 represents the most challenging grades for chess teacher since there is abundant chess knowledge related to math but is not matched by math curriculum learning outcomes from grades K to 4. They are, though, matched at higher grades from grade 5 to grade 7. However, when compared to higher grades, for the same reason, these students from kindergarten to grade 4 are also the students who would benefit the most by learning chess with a higher possibility of improving their math scores. The chess skills learned, including algebraic notation, checkmate pattern and tactics pattern, how the chess pieces move, and their respective values, could be transferred to math concepts.

Student's name: ________________ Assignment date: ____________________

By working on mathematical chess puzzles, students get training on how to transfer chess knowledge to improve math ability. Since chess is a whole number based strategy game, so it is important for students to get exposure to computational mathematical chess puzzles. Examples of mathematical chess puzzles incorporating math-learning outcomes from Grades K to grade 7 are included in this article.

Algebraic notation learned in chess could be transferred to the concept of coordinates, which are not introduced until grade 6. For this reason, lots of attention and practices must be given to students who are lower in grade 6. When compared math learning outcomes and math knowledge learned in chess, chess is related to number operations, number concepts, but students would benefit more if they are given the opportunities to work on integrated math and chess problems.

Chess patterns are very different from the math pattern problems in the sense that the chess pattern has a cause-consequence effect and does not use the relation of the adjacent terms to predict the next term. The chess tactics and checkmate patterns use the special formation to jointly produce its result(s). The most interesting is one would not learn this kind of pattern other than by playing chess.

The king's triangular shape of movement to create opposition in chess is an example of how the use of a geometric shape would take a special meaning in chess. How about the distance in knight moves? It may take fewer moves for a knight to reach a faraway square than to get to a nearby square. The diagonal distance is the same as the side length of a square when promoting a pawn; all these are very intriguing when thinking from math's point of view. These have to be clearly explained to elementary students. See examples in this article for more explanations.

One notable math concept learned in playing chess but not widely taught is the set theory. Chess players constantly use the concept of the Venn diagram to look for interactions among chess pieces. The chess game itself is highly related to data gathering and information analysis. Its relation to statistics and probability could be highlighted by working on math and chess puzzles.

Student's name: ________________ Assignment date: ____________________

The effect of transferring math knowledge learned in chess will be less significant if the chess teacher does not take the efforts to emphatically point out the math concepts. The task of transferring math knowledge learned in playing chess would be much easier if students are offered the opportunities to work on mathematical chess puzzles.

Many children could not play chess well but feel very proud that they could solve mathematical chess puzzles. Mathematical chess puzzles provide some children with additional opportunities that they could challenge themselves. For this reason, I give prizes to chess winners and also to puzzles solvers.

The most interesting about using chess symbols is that the chess symbols themselves not only possess pre-defined values but also have the implied meaning of movements, and these two special characters allow me to create some very interesting mathematical puzzles with pizzazz.

By using chess symbols, a simple one-step arithmetic problem could become a multi-step problem. As a result, chess symbols and values offer children more opportunities to work on other types of questions, which could stimulate children's brain cells and improve their problem-solving ability. So the benefits of working on these types of problems are double-edged- improving chess knowledge and also mathematical problem-solving ability.

Student's name: ________________ Assignment date: ____________________

Example 1 - Number Concepts and Operations: Addition and Subtraction

The following problem is designed to be different from traditional worksheets, which are always from left to right or top to down in a linear fashion. One could work out the problem below from the bottom to top and then from top to down in multi-direction. It uses a chess symbol as part of the computation. Thus multi-step problems are created.

Bottom/up and top/down operation	Bottom/up and top/down operation
□ ; ———— ; + ♝ ; ♛ 12 ; − ♝ ; ———— ; □	□ ; ———— ; + ♜ ; 6 □ ; − ♜ ; ———— ; 6

Student's name: ________________ Assignment date: ____________________

Example 2 - Number Concepts and Operations: Multiplication of doubling

The following problem is created with the mind that children do not really learn math in a sequential way of addition, subtraction, multiplication, or division in real life. This example demonstrates times table created using different formats. A simple multiplication problem is changed to a problem in multi-direction, multi-operation, multi-step, and multi-concept learning.

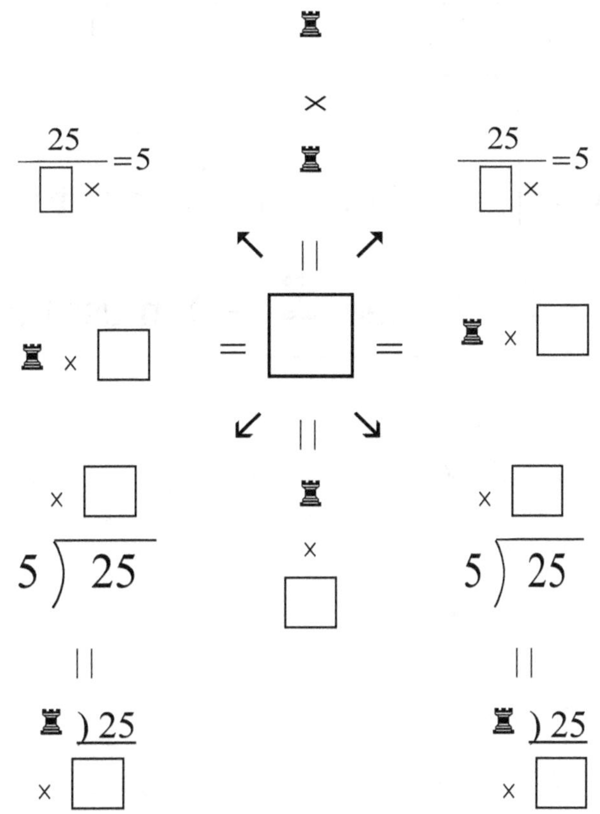

Student's name: ________________ Assignment date: ____________________

Example 3 - Number Concepts and Operations: Addition and Subtraction, If Then - Else

The following operation takes a circular motion in clockwise.

	19	−	♚	=	□	
	9	+	♚	=	□	+
+	□				□	

If 10 + ♜ = □, then 9 + ♜ must be □.

If ♜ + 10 = □, then ♜ + 9 must be □.

Student's name: ________________ Assignment date: ____________________

Example 4 - Number Concepts and Operations: Cross Multiplication

A five-step question could be used as a method to do factoring in grade 9.

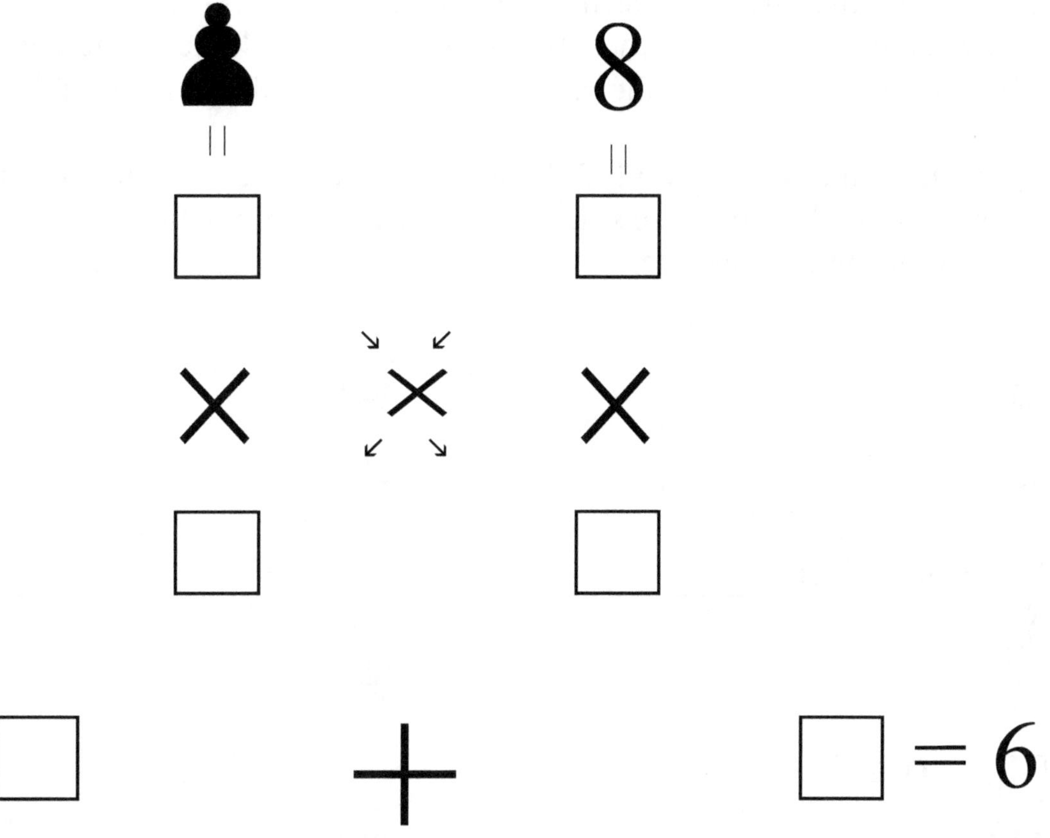

Student's name: ________________ Assignment date: ____________________

Example 5 - Number Concepts and Operations: Multiplication and Division

Will the following operations confuse children? The answer is no. ♛×♞ will not make sense if it is explained literally as a queen times a knight. However, if it is translated into numerals 9 times 3, then the product must be 27, which is very logical, and children understand that they are working on the product of 9 × 5, not the product of ♛×♞.

A similar logic question is as follows: If 2 # 3 is defined as 2 + 2 × 3, then what is 3 # 4? Usually, 2 # 3 will not make any sense since it is not a valid arithmetic operator, but if we define it clearly, it becomes workable.

♛ × 2 18 ÷ 2 = □	♛ × ♜ □ ÷ 5 = □
♛ × ♜ □ ÷ 9 = □	♛ × ♛ □ ÷ ♛ = □
♛ × ♞ □ ÷ ♞ = □	♛ × ♞ □ ÷ ♛ = □

Student's name: ________________ Assignment date: ___________________

Example 6 – Pattern and Relations: Equation

The following example demonstrates how chess symbols and chess values are integrated with arithmetic operations.

♛ + ♞ + x = 54

x = _____

Example 7 – Pattern and Relations: Pattern

Chess pieces can move horizontally, vertically or diagonally, and the concept of symmetry is not in the learning outcome of grade K to 1, so they must thoroughly be explained to students. An example of a math and chess integrated puzzle using a chess move is as follows.

Use chess moves to solve the following puzzle.

On the first look, lots of students are not able to solve it. Why? Students are so used to do computation from left to right, and this question has to be solved in an unconventional direction. Chess is a 360^0 Visualization game, and this example demonstrates how a knight move would help solve this puzzle.

	?		14	
?				21
		♘		
?				28
	42		?	

Student's name: ________________ Assignment date: ____________________

Example 8 Set Theory shown by two-column format

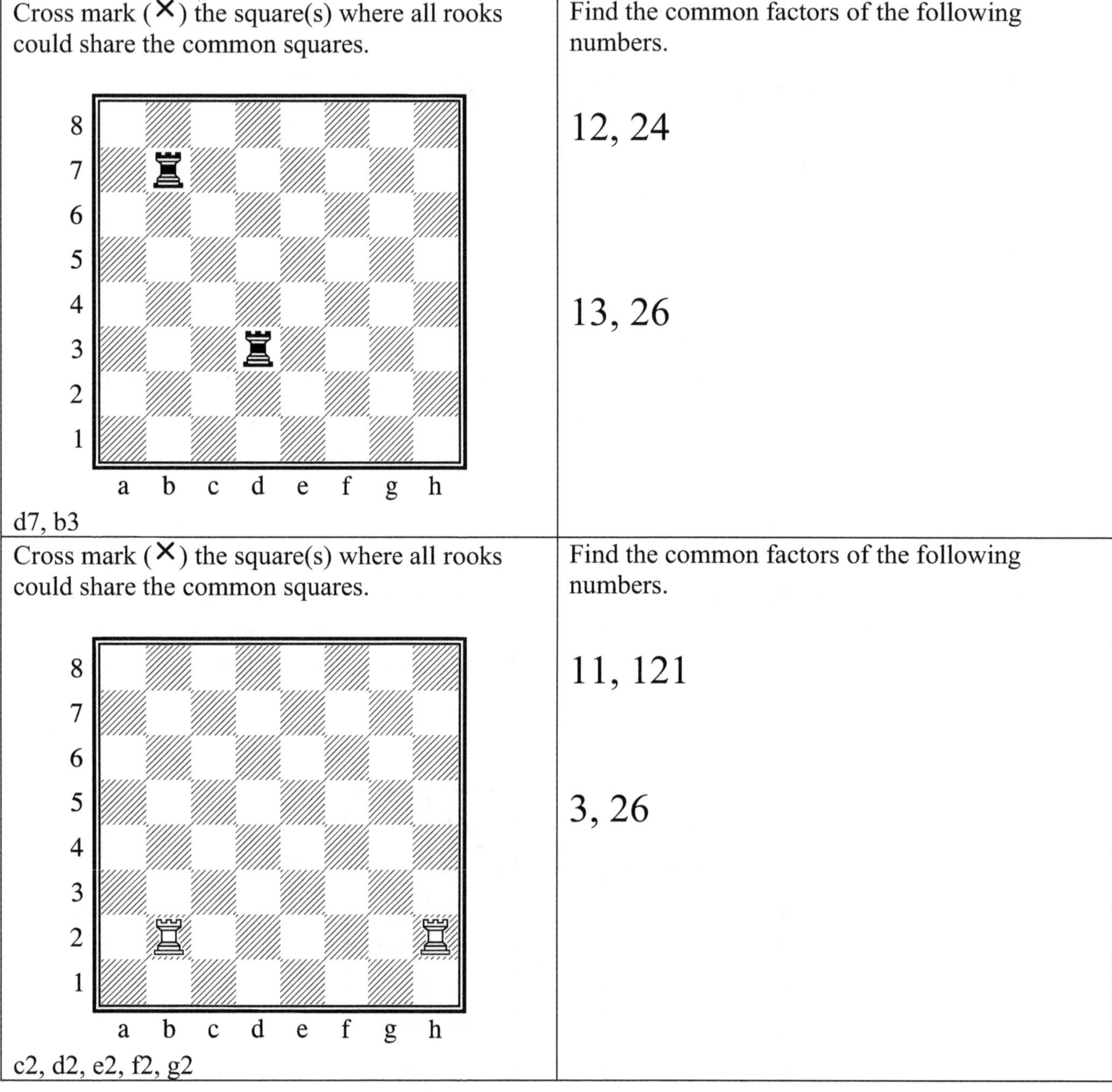

Cross mark (✕) the square(s) where all rooks could share the common squares. d7, b3	Find the common factors of the following numbers. 12, 24 13, 26
Cross mark (✕) the square(s) where all rooks could share the common squares. c2, d2, e2, f2, g2	Find the common factors of the following numbers. 11, 121 3, 26

Student's name: ________________ Assignment date: ____________________

Example 9 An example using chess moves

If ♛ ÷ ♝ = ♜
Then what is ♚ ÷ ♜ = ?

Example 10 Shape and Space - The following is a puzzle that requires the knowledge of chess moves.

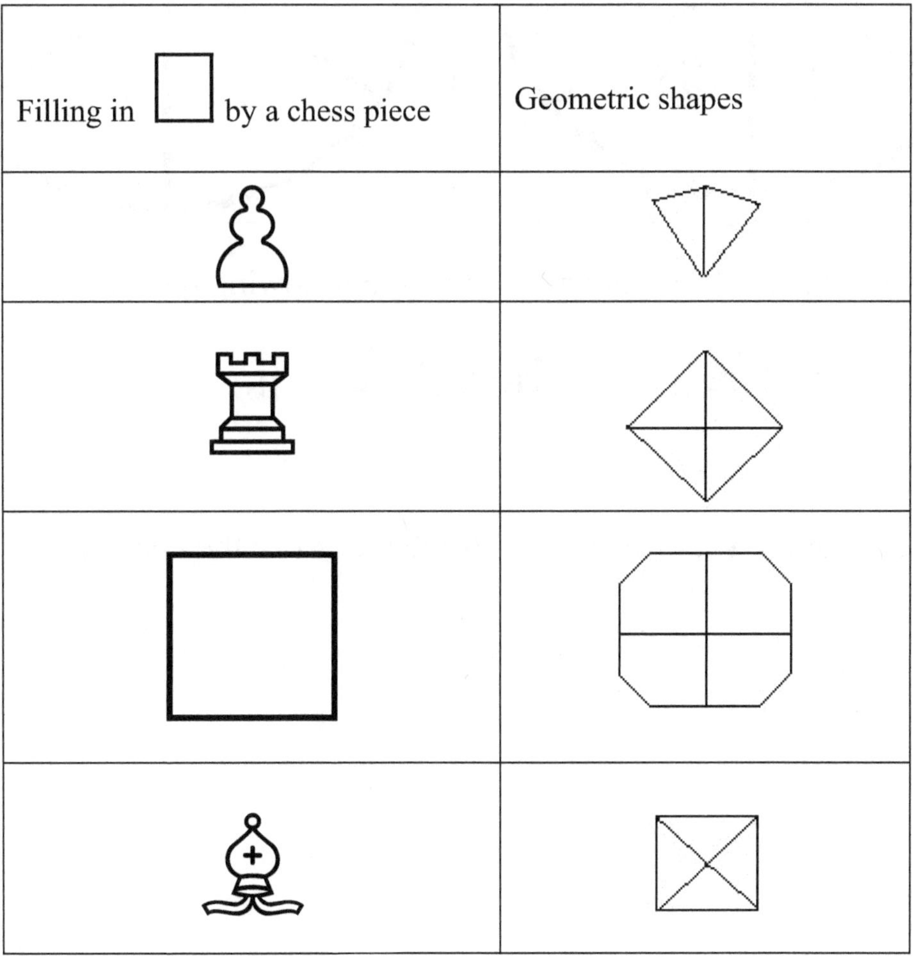

Student's name: ________________ Assignment date: ____________________

Example 11 – Pattern and Relations

Find values to replace? or fill in □. An example using chess pieces values and logic is as follows.

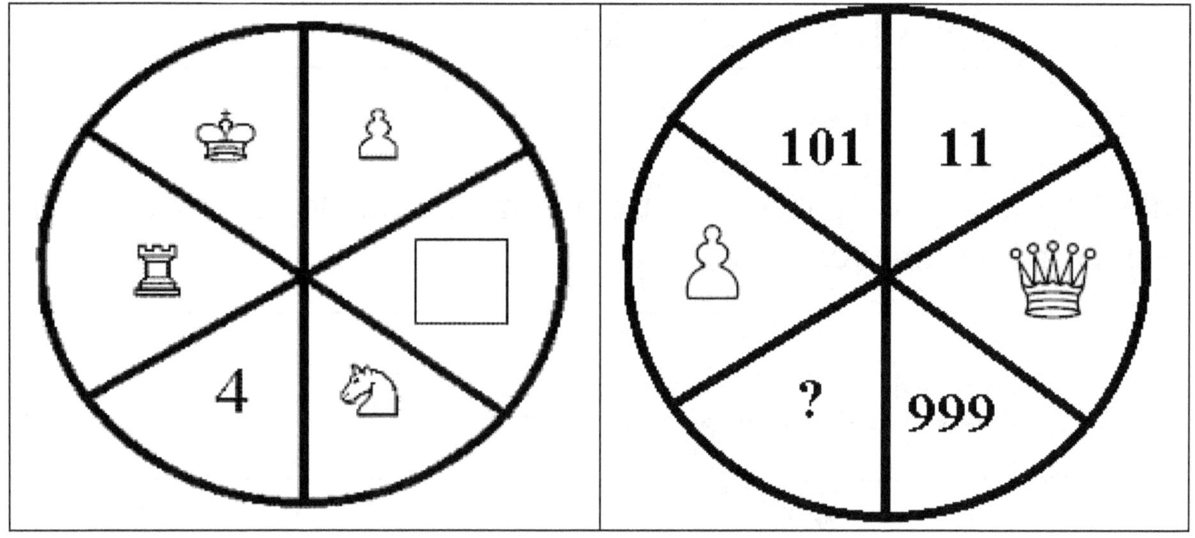

Example 12 - Use chess symbol moves to solve the following puzzle.

If 2 ♜ 3 = 5 then 2 ♝ 3 is = _____

Surprisingly, some of my students have no trouble to solve the above puzzle.

Student's name: ________________ Assignment date: ___________________

Example 13 – Shape and Space

Use chess symbol moves to solve the following puzzle.

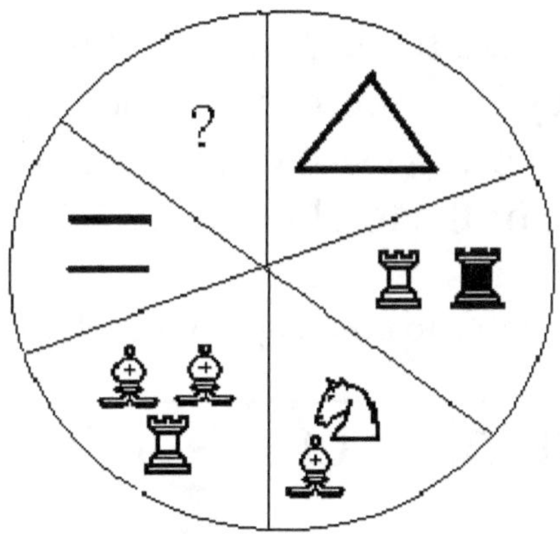

Student's name: ________________ Assignment date: ____________________

Example 14 - Statistics and Probability

Table Values

The use of chess values is much like the use of monetary values. When chess or money figures are seen by children, they both represent some pre-defined meaningful values. The following is an example where the values of chess pieces could be monetary values and "Total Points" could be the sum of the total money.

Fill in the different number of chess pieces to come up with each total.

Number of ♟	Number of ♞	Number of ♜	Total points
1	1	1	9
3	2	0	9
0	3	0	9
☐	☐	☐	10
☐	☐	☐	10
☐	☐	☐	11
☐	☐	☐	12
☐	☐	☐	13
☐	☐	☐	14
☐	☐	☐	15

Student's name: ________________ Assignment date: ____________________

Example 15 - Statistics and Probability

Chess Puzzle Samples	Expected Math Learning Outcomes	Chess knowledge required
What is the probability of the following two pieces of meeting together? 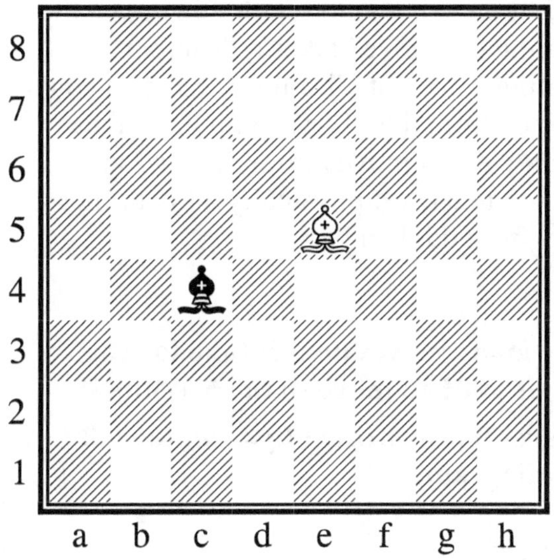 [Chessboard diagram: black bishop on c4, white bishop on e5; ranks 8–1, files a b c d e f g h]	▪ Probability	▪ Chess pieces moves

Why buy a Ho Math Chess Franchise?

Intellectual Properties of Ho Math Chess

何数棋谜 - www.homathchess.com

Student's name: ________________ Assignment date: ____________________

Example 16 - How does the knight move?

How does a knight move?	When 2 is not = 2?
Move the knight at d5 to each square identified by ⊡. For example, it will take 2 moves from d5 to move to b5. Nd5 – c7 – b5. Write the least number of moves required to reach each ⊡ from Nd5 on the squares of ⊡ .	**Observe your results and see if all your answers are 2's?** **Answer: ________________.** If each move is identified as 1 and 2 moves are 2, then you would see that in chess, the meaning of "move" takes a different meaning since not all 2-move have an equal distance from d 5. For example, it takes knight 2 equal moves from d5 to b5 and from d5 to a4, yet physically a4 is farther away from d5 than b5. **How many moves does it take for the knight at d5 to move to the 5 White squares not identified by ⊡ on the left side diagram?** **Answer: ______________________**

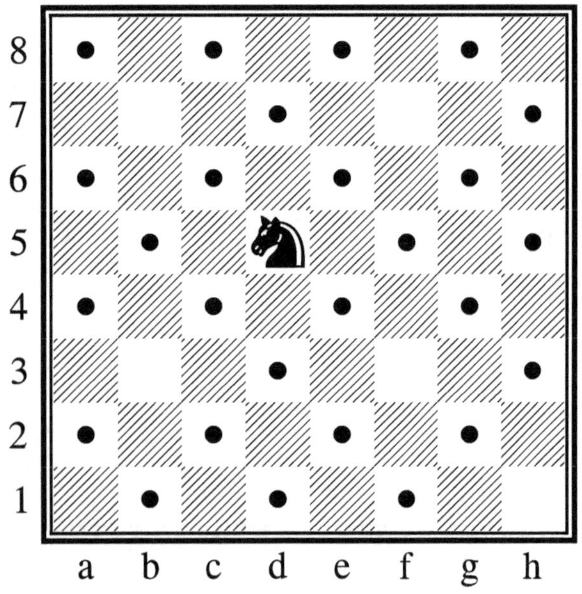

Yes, 4 moves

Student's name: ________________ Assignment date: ____________________

Example 17 Diagonal length = Side length

Is the distance from c6 to c8 the same as distance c6 to a8?

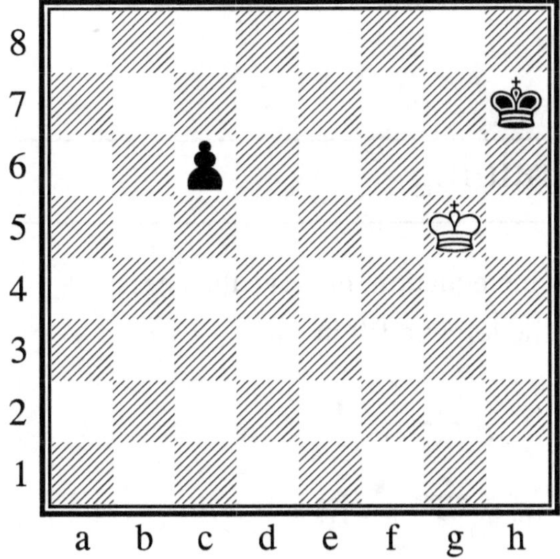

Student's name: ________________ Assignment date: ________________

Example 18 - Pattern and Relations

The following problem demonstrates that not only children will not get confused about traditional chess symbols used in an arithmetic expression; they could be led to use additional "creative" chess symbols and are able to solve them correctly. This problem is suitable for grade 3 and above students.

Chess Symbol	Logic Training
A new Chess Symbol is defined as follows: Chess figurines / Chess symbols: ♔ — ÷ (Opposition) ♖ — ┼ ♘ — L ♗ — X ♕ — ✳ ♙ — ∨ with a vertical stroke	In the following equation, observe the chess symbols on the left and fill in each ◯ with a number. If ┼ + ┼ = 10 then ÷ + ┼ = ◯ ┼ ✳ + ┼ ÷ ――――― ◯ ÷ ✳

Use the above Chess Symbol table; find the following pattern:

Z, ÷, O, ∨, T, L, T, X, F, ┼, ______, ✳

Use the above Chess Symbol table; find the following pattern:

0, ÷, 1, ∨, ____, L, 3, X, 5, ┼, ____, ✳

Student's name: ________________ Assignment date: ___________________

Reference

(1) Chess & Math Add Up, Yee Wang Fung, Teacher, May/June 1995
(2) Chess and Standardized Test Scores, James Liptrap, Chess Life, March 19998
(3) Chess curriculum aligned with the National Mathematics Standards, unpublished manuscript, John Buky, 2005

Student's name: ________________ Assignment date: ____________________

Appendix A

Comparison (Based on 2005 Canada BC school curriculum)

Number Concepts and Operations

Grades	Math Learning Outcomes	Matched chess knowledge to the math learning outcomes	Math knowledge learned not matched by chess	Math knowledge learned in chess but not included in math curriculum
Grades K to 1	Recognize, describe, and use numbers from 0 to 100 in a variety of familiar settings. Demonstrate and use a variety of methods to show the process of addition and subtraction on one-digit whole numbers.	• Chess pieces values of 0, 1, 3, 5, and 9 • Counting chess pieces values • Compare object with values • Understand the concept of "half" of the chessboard	• Skip count to 100 by 1s, 2s, 5s, and 10s. • Estimating and Comparing estimates • Addition and subtraction to 18.	• Algebraic notation • Cancellation of equal values of chess pieces. • Counting in multi-direction with multiple attacks • Special tactics pattern • Chess pieces movements • Checkmate pattern • Attacking sequence • Interaction square • Logic
Grades 2 to 3	Develop a number sense for whole numbers from 0 to 1000 and common fractions to tenths. Use a variety of strategies to apply a basic operation (+, −, ×, ÷) to whole numbers and use these operations in solving problems.	• Chess pieces values of 0, 1, 3, 5, and 9 • Counting chess pieces values • Compare object with values • Understand the concept of "half" of the chessboard	• Estimating Rounding to the nearest 10 and 100 • Skip count forward and backward by 2s, 5s, 10s, 25s, and 100s to 1000. • Writing number in words • Place values • Divisibility by 2, 5, 10 • Multiplication up to 25 (5 × 5) • Even and odd • Understanding of halves, thirds, fourths, fifths, and tenths.	• Algebraic notation • Cancellation of equal values of chess pieces. • Counting in multi-direction with multiple attacks • Special tactics pattern • Chess pieces movements • Checkmate pattern • Attacking sequence • Interaction square • Logic

Student's name: ________________ Assignment date: __________________

Comparison

Number Concepts and Operations

Grades	Math Learning Outcomes	Matched chess knowledge to the math learning outcomes	Math knowledge learned not matched by chess	Math knowledge learned in chess but not included in math curriculum
Grade 4	Demonstrate a number sense for whole numbers from 0 to 10000 and for proper fractions. Apply arithmetic operations on whole numbers and illustrate their use in solving problems.	• Chess pieces values of 0, 1, 3, 5, and 9 • Counting chess pieces values • Compare object with values	• Estimating • Rounding to the nearest 10, 100, and 1000 • Skip counting • Writing number in words • Connect proper fractions to decimal fractions (tenths and hundredths) • Multiplication up to 9×9	• Algebraic notation • Cancellation of equal values of chess pieces. • Counting in multi-direction with multiple attacks • Special tactics pattern • Chess pieces movements • Checkmate pattern • Attacking sequence • Interaction square • Logic
Grade 5	Demonstrate a number sense for whole numbers, from 0 to 100000, and will explore proper fractions and decimal fractions. Apply arithmetic operations on whole numbers and decimal fractions and illustrate the use of decimal fractions in solving problems.	• Chess pieces values of 0, 1, 3, 5, and 9 • Counting chess pieces values • Compare object with values	• Place value from hundredths • Read and write numerals to a million • Estimate up to 100000 • Recognize multiples, factors, composites, and primes • Describe proper fractions, equivalent fractions • ddd × dd • ddd ÷ d • dd.dd × d • dd.dd ÷ d	• Algebraic notation • Cancellation of equal values of chess pieces. • Counting in multi-direction with multiple attacks • Special tactics pattern • Chess pieces movements • Checkmate pattern • Attacking sequence • Interaction square • Logic

Student's name: ________________ Assignment date: ____________________

Comparison

Number Concepts and Operations

Grades	Math Learning Outcomes	Matched chess knowledge to the math learning outcomes	Math knowledge learned not matched by chess	Math knowledge learned in chess but not included in math curriculum
Grade 6	Develop a number sense for common fractions and explore number sense for whole numbers. Apply arithmetic operations on whole numbers and decimal fractions in solving problems.	• Chess pieces values of 0, 1, 3, 5, and 9 • Counting chess pieces values • Compare object with values	• Read and write numerals > a million • Multiples, factors, composites, and primes • Exponent • Integers • Compare fractions, mixed numbers, and decimal fractions • Ratio and percentage	• Algebraic notation • Cancellation of equal values of chess pieces. • Counting in multi-direction with multiple attacks • Special tactics pattern • Chess pieces movements • Checkmate pattern • Attacking sequence • Interaction square • Logic
Grade 7	Demonstrate a number sense for decimal fractions and integers (including whole numbers). Apply arithmetic operations on decimal fractions and integers and illustrate their use in solving problems.	• Chess pieces values of 0, 1, 3, 5, and 9 • Counting chess pieces values • Compare object with values • Set theory for common factors • The decision for prime factorization • Attacking sequence for the order of operations	• Common multiples, common factors, least common multiples, greatest common factors, and prime factorization • Order of operations • Scientific notation • Divisibility rules by 2, 3, 4, 5, 6, 8, 9, 10, or 11 • Expand standard numbers in posers of 10. • All 4 operations (+, −, ×, ÷) of decimal fractions. • Convert proper fractions, decimal fractions, terminating and repeating decimals, and % from one to another. • Rate, ratio, and proportion • ddd.dd × dd • ddd.dd ÷ dd • ddd.dd × ddd • ddd.dd ÷ ddd • All 4 operations (+, −, ×, ÷) of integers.	• Algebraic notation • Cancellation of equal values of chess pieces. • Counting in multi-direction with multiple attacks • Special tactics pattern • Chess pieces movements • Checkmate pattern • Attacking sequence • Interaction square • Logic

Student's name: ________________ Assignment date: ___________________

Comparison

Pattern and Relations (Patterns, Variable and Equations (grade 6 and above only), Measurement

Grades	Math Learning Outcomes	Matched chess knowledge to the math learning outcomes	Math knowledge learned not matched by chess	Math knowledge learned in chess but not included in math curriculum
Grades K to 1	Identify, create, and compare patterns that arise from their daily experiences. Estimate, measure, and compare measures using whole numbers and non-standard units of measure.		• Compare length, size, area, weight, and volume. • Describe time and temperature	• The real distance and chess distance such as knight's move and promotion square and triangular moves of opposition.
Grade 2 to 3	Investigate, establish, and communicate rules that arise from daily and mathematical experiences, and use these rules to make predictions. Measure, estimate, and compare, using whole numbers and non-standard and standard units of measure.		• Measure length, mass, volume, and time. • Read and write time to the nearest minute using 12-hour notation. • Estimate, read, and record temperature to the nearest degree Celsius. • Count, write and read money.	• The real distance and chess distance such as knight's move and promotion square and triangular moves of opposition.
Grade 4	Investigate, establish, and communicate rules and predictions from numerical and non-numerical patterns. Estimate, measure, and compare quantities using numbers and standard units of measure.	• Algebraic notation	• Use grids, tables, charts, or calculators to explain mathematical relationships. • Measure objects in length, area, and capacity, mass. • Read and write time on a 24-hour clock and a.m., p.m.	• The real distance and chess distance such as knight's move and promotion square and triangular moves of opposition.

Student's name: ________________ Assignment date: ____________________

Comparison

Pattern and Relations (Patterns, Variable and Equations (grade 6 and above only), Measurement

Grades	Math Learning Outcomes	Matched chess knowledge to the math learning outcomes	Math knowledge learned not matched by chess	Math knowledge learned in chess but not included in math curriculum
Grade 5	Construct, extend, and summarize patterns using rules, charts, mental mathematics, and calculators. Use measurement concepts, appropriate tools, and the results of measurements to solve problems in real-life contexts.		• Construct patterns in two or three-dimension. • Explain the meaning of length, width, height, depth, thickness, perimeter, and circumference. • Solve problems in mass, perimeter, area, volume.	• Special tactics pattern • Chess pieces movements • Checkmate pattern
Grade 6	Use relationships to summarize, generalize, and extend patterns. Use informal and concrete representations of equality and operations on equality to solve problems. Be able to solve problems involving perimeter, area, surface area, volume, and angle measurement.	• Equal points of attacking pieces and defending pieces • Chess pieces points substitute chess symbols	• Solve one-variable equations with whole-number coefficients. • Create expressions and rules to describe patterns and relationships of area, perimeter, volume etc. • Determine the perimeter of polygons. • Estimate angles using a circular protractor. • Classify angles as acute, right, obtuse, straight, or reflex.	• Special tactics pattern • Chess pieces movements • Checkmate pattern • Inequality points of attacking pieces and defending pieces
Grade 7	Express patterns in terms of variables and use expressions containing variables to make predictions. Use variables and equations to express, summarize, and apply relationships as problem-solving tools in a restricted range of contexts. Solve problems involving the properties of circles and their relationships to angles and time zones.	• Equal points of attacking pieces and defending pieces • Chess pieces points substitute chess symbols • Angles formed by the movements of chess pieces.	• Solve and verify simple linear equations. • Substitute number variables • Write expressions. • Predict and justify the nth value of a number pattern. • Calculate time in different time zones. • Solve circumference and area of a circle.	• Special tactics pattern • Chess pieces movements • Checkmate pattern • Inequality points of attacking pieces and defending pieces • Parallel and intersection of chess pieces.

Student's name: ________________ Assignment date: ____________________

Comparison

Shape and Space (3-D Objects and 2-D Shapes, Transformations)

Grades	Math Learning Outcomes	Matched chess knowledge to the math learning outcomes	Math knowledge learned not matched by chess	Math knowledge learned in chess but not included in math curriculum
Grades K to 1	Explore, sort, and classify real-world and three-dimensioned objects according to their properties.	• Square-shaped chessboard	• Identify circles, squares, triangles, or rectangles. • Describe reflection	• Pattern of chessboard • Lines, shapes, and patterns of the chess pieces on the chessboard.
Grade 2 to 3	Describe, classify, construct, and relate three-dimensional objects and two-dimensional shapes using common language to describe their properties.	• Bishop vs. parallel, checkmate vs. intersecting, rook vs. perpendicular.	• Describe faces, vertices, edges, sides, and angles of polygon and solids. • Name three-dimensional objects cubes, spheres, cones, cylinders, pyramids, and prisms. • Explore the concepts of points, lines, perpendicular lines, parallel lines, and intersecting lines on three-dimensional objects. • Graph whole numbers on horizontal or vertical number lines.	• Chess pieces move in 360 degrees. • Chess coordinates. • Lines, shapes, and patterns of the chess pieces on the chessboard.
Grade 4	Describe, classify, construct, and relate three-dimensional objects and two-dimensional shapes using mathematical vocabulary to describe their properties.	• Bishop vs. parallel, checkmate vs. intersecting, rook vs. perpendicular. • Algebraic notation.	• Construct nets for pyramids and prisms. • Identify squares, rectangles, parallelograms, and trapezoids. • Draw point, line, parallel lines, and intersecting lines. • Place an object on a grid using columns and rows.	• Chess pieces move in 360 degrees. • Lines, shapes, and patterns of the chess pieces on the chessboard.

Student's name: ________________ Assignment date: ____________________

Comparison
Shape and Space (3-D Objects and 2-D Shapes, Transformations)

Grades	Math Learning Outcomes	Matched chess knowledge to the math learning outcomes	Math knowledge learned not matched by chess	Math knowledge learned in chess but not included in math curriculum
Grade 5	Use the visualization of two-dimensional shapes and three-dimensional objects to solve problems related to spatial relationships. Describe motion in terms of a slide, a turn, or a flip.	Chess coordinates – first quadrant only.	• Construct patterns in two or three-dimension. • Explain the meaning of length, width, height, depth, thickness, perimeter, and circumference. • Solve problems in mass, perimeter, area, volume. • Classify triangles according to sides. • Classify polygons according to sides of 3, 4, 5, 6, and 8. • Recognize translation, turn, or a flip, and tessellations • Identify the point in the first quadrant using order pairs.	• The triangular move of opposition • Movement of the chess pieces on a 2-dimensional plane: how the Knight makes its tour on the chessboard touching each square only once. • In chess, time, space and material interact in a "dynamic" or "flux" to create imbalances.
Grade 6	Use visualization and symmetry to solve problems involving classification and sketching. Create patterns and designs that incorporate symmetry, translations, tessellations, and reflections.	Chess coordinates – first quadrant only.	• Classify angles as acute, right, obtuse, straight, or reflex. • Classify triangles according to angles. • Regular polygons. • Draw solids on grids. • Draw slides, flips and turns using grids to describe.	• The triangular move of opposition • Movement of the chess pieces on a 2-dimensional plane: how the Knight makes its tour on the chessboard touching each square only once. • In chess, time, space and material interact in a "dynamic" or "flux" to create imbalances.
Grade 7	Link angle measurement to the properties of parallel lines. Create and analyze patterns and designs using congruence, symmetry, translation, rotation, and reflection.	Chess coordinates – first quadrant only.	• Complementary and supplementary angles. • Solve circumference and area of a circle. • Angles related two parallel lines. • Construct angle bisectors and perpendicular bisectors. • Draw slides, flips and turns using grids to describe in all quadrants.	• The triangular move of opposition • Movement of the chess pieces on a 2-dimensional plane: how the Knight makes its tour on the chessboard touching each square only once. • In chess, time, space and material interact in a "dynamic" or "flux" to create imbalances.

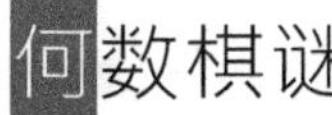

Student's name: _______________ Assignment date: _________________

Comparison

Statistics and probability (Data Analysis, Chance, and Uncertainty)

Grades	Math Learning Outcomes	Matched chess knowledge to the math learning outcomes	Math knowledge learned not matched by chess	Math knowledge learned in chess but not included in math curriculum
Grades K to 1	Collect, organize, and analyze (with assistance) data based on first-hand information. Predict the chance of an event happening using the terms never, sometimes, and always.	• Chessboard • Make the next move using tree structure	• Construct a pictograph • Pose oral questions in relation to the data gathered.	• Number of possibilities of making the 1st move in chess.
Grade 2 to 3	Collect data based on first-and-second-hand information, display results in more than one way, interpret data and make predictions. Use simple experiments designed by others to illustrate and explain probability and chance.	• Chessboard	• Display data in graphs, pictographs, bar graphs, and rank ordering. • Obtain information by performing arithmetic operations on the data. • Describe an outcome in terms such as likely, unlikely, fair chance, probable, and expected. • Conduct a probability experiment.	• Number of possibilities of making the 1st move in chess.
Grade 4	Collect first-and second-hand data, assess and validate the data-collection process, and graph the data. Conduct simple probability experiments to explain outcomes.	• Chessboard	• Construct bar graph and pictograph using many-to-one correspondence. • Identify an outcome in terms of possible, impossible, certain, uncertain. • Compare outcomes using terms equally, likely, more likely, or less likely.	• Number of possibilities of making the 1st move in chess.

Student's name: ________________ Assignment date: ____________________

Comparison

Statistics and probability (Data Analysis, Chance, and Uncertainty)

Grades	Math Learning Outcomes	Matched chess knowledge to the math learning outcomes	Math knowledge learned not matched by chess	Math knowledge learned in chess but not included in math curriculum
Grade 5	Develop and implement a plan for the collection, display, and analysis of data gathered from appropriate samples. Predict outcomes, conduct experiments, and communicate the probability of single events.		• Distinguish between a population and sample. • Display frequency diagram, line plot, and broken-line plot. • Make inferences from the data to generate a conclusion. • Sample space • Use terms never/less likely/equally likely/more likely/always • Conduct probability experiments.	• Statistical analysis of opening moves that lead to wins for the white or black pieces. • Analysis of a given position in a chess database to evaluate possible moves. • Analyze combination of moves up to move 10.
Grade 6	Develop and implement a plan for the collection, display, and analysis of data gathered from appropriate samples. Use numbers to communicate the probability of single events from experiments and models.		• Justify sampling techniques. • Display data in histogram, double bar graphs, and stem and leaf plot. • Describe minimum, maximum, mode, median, and patterns. • Dice problems. • Distinguish experimental and theoretical probability of single events.	• Statistical analysis of opening moves that lead to wins for the white or black pieces. • Analysis of a given position in a chess database to evaluate possible moves. • Analyze combination of moves up to move 10.
Grade 7	Develop and implement a plan for the collection, display, and analysis of data, using measures of variability and central tendency. Create and solve problems using Probability.		• Select and justify appropriate methods of collecting data. • Calculate mode, median, mean, range, extremes, gaps, quartiles, and clusters. • Interpolate from data to make predictions • Solve problems using the definition of probability as favourable outcomes over total outcomes.	• Statistical analysis of opening moves that lead to wins for the white or black pieces. • Analysis of a given position in a chess database to evaluate possible moves. • Analyze combination of moves up to move 10.

Student's name: ________________ Assignment date: ____________________

References

(1) Chess & Math Ad Up, Yee Wang Fung, Teach, May/June 1995. p. 15

(2) Chess and Standardized Test Scores, James Liptrap, Chess Life, March 1998. pp 41-43

Student's name: ________________ Assignment date: ____________________

Comparing Kumon Math to Ho Math Chess

Many have asked me how Ho Math Chess is different from Kumon Math. Both Kumon Math and Ho Math Chess use worksheets to teach math and get children to work on progressive math worksheets. The biggest difference is the way the worksheets are produced.

Ho Math Chess teaching idea stems from its founder's Mr. Frank Ho's personal experience while teaching his son chess when he was 5 years old. This is very similar to Kumon math's founder Mr. Toru Kumon's experience in teaching his son math. Both Mr. Kumon and Mr. Ho have gained insights by teaching their own child in creating the franchise business in teaching math but in a very different approach of creating worksheets.

Ho Math Chess was founded on the philosophy that math must be taught in a fun and cool way. Because of this reason, Ho Math Chess integrates game and math together and the game used at Ho Math Chess is the international chess – a universal language which many kids in the world can play and enjoy while working on math.

Mr. Ho discovered the language which links between chess and math and thus patented his idea and created the world's first math and chess integrated workbooks for elementary students. Consequently, Mr. Ho also created Ho Math Chess Teaching Set so that children as young as 4 years old can learn chess with ease. Frank also created Frankho Chess Mazes, which is very different from traditional mazes in such a way that Frankho Chess Mazes improve children's visualization ability and also their analytical ability.

The math worksheets at Ho Math Chess integrate math and chess using new and patented applied technology with innovative teaching idea.

For children who do not just want to work on pure math worksheets only then, Ho Math Chess offers an alternative. Mr. Ho also created mathematical chess puzzles, which resemble IQ puzzles so children can have the opportunity to do brain fitness and improve their brain power at the same time. Ho Math Chess is taking the leading role of integrating math, chess, puzzles, and brain fitness all into one workbook. Ho Math Chess worksheets are fun, entertaining, and educational.

It is the teaching philosophy of having fun while learning math differs Ho Math Chess from Kumon math and all other math learning centers in the world.

More details, please visit www.mathandchess.com.

Student's name: ________________ Assignment date: ___________________

Comparing Traditional Chess Set to the Flat-Faced Chess Set

It is interesting to note that western chess set are all made in 3-D figurines and each chess piece looks very different and yet Chinese chess sets are all flat-faced with uniform look on one surface, why is this? I do not know if anyone has done a research to look into and see why there is such a disparity in using such a very different style?

I have invented a new chess training set which is based on geometry lines in correspondence to how each chess piece moves. Since the emphasis is so much on its direction of moves so I used the flat-surfaced chess pieces. After I used it for a while, I concluded that my flat-surfaced international chess set is just as effective to use when compared to the traditional chess set, but I am not asking adults to convert to use my chess set. The clear advantage of using my invention - Ho Math Chess Teaching set is that it is so easy for young children to learn chess and the speed that the young children can learn to play when using Ho Math Chess Teaching Set is just incredible. This can be evident by the responses from the accompanying parents who come to my class with kids together and asked for using Ho Math Chess Teaching Set since they also think it is much easier for their children to learn how to play chess.

The following attempts to make some comparisons on how different there are between the traditional chess set and Ho Math Chess Teaching Set (hereby called Ho Set).

1. Look

Ho Set is flat-surfaced, and each chess piece's moves are marked on the surface of each chess piece, so it is easier for young children to learn how to move by just looking at the directions marked on each chess piece when compared to the use of traditional chess set.

2. Geometry Concept

   Ho Set design is based on the geometry concept of line and line segment to indicate that the allowable move is one square only or multi-square. Each arrow marked on each chess piece metaphors a line so the move could be multi-square, whereas the line segment with no arrows signals the move is just one square at a time. The traditional set's figurines have no relation to its move directions. The Ho Set's move direction stems from the middle of each squared chess piece (turn point for knight), and thus the lines are actual symmetry lines of each square. The entire design of the Ho Set is closely tied with the concept of geometry, but the traditional chess set is not.

3. Blind chess and half-blind chess

Student's name: ________________ Assignment date: ____________________

Since the flat-surfaced Ho Set has a uniform look on one side, so it allows children to play blind chess or half-blind chess. The half-blind chess has 3 advantages: one is that it does not require players to be blindfolded and is hilariously fun to play. The third advantage is it trains children's memory since one has to remember where the opponent's chess pieces are.

4. Incorporating with workbooks

   The Ho Set symbols are built into Ho Math Chess integrated workbooks, so the learning curve in the use of the math and chess integrated workbook is shortened by using Ho Set.

As it can be easily concluded that Ho Set is very good at training children on how to play chess. In addition, it can also be used in conjunction with working on Ho Math Chess integrated workbooks.

Student's name: ________________ Assignment date: ____________________

Enrichment or gifted mathematics teaching for gifted

The development of a gifted math program at a commercial learning centre is very different from the enrichment or gifted program at regular day school. In this article, we like to share our experience on how we at Ho Math Chess develop or create our own gifted or enrichment program. When creating a gifted math program at the learning centre, the learning centre must be aware of the following important factors.

Why gifted math program at the learning centre is different

The math program at Ho math Chess normally runs in the length of 2 hours, for very young kids, their attention span will not last for 2 hours, so at the learning centre, we must develop a gifted math program which can get children's attention for a longer time, much longer than school math allotted teaching time. For this reason, the material must be fun and entertaining.

Math materials must be multi-level and be able to teach students with mixed-ability

It is very normal, we have different grades in one class, and even all at the same grades but their math abilities will be different since they come from different day schools. So our math teaching materials must be able to handle multi-level and are suitable for students with different math backgrounds.

Everybody needs to be challenged

It is very wrong to assume that students with abilities less than gifted children do not need to be challenged; they need to be challenged at different levels. Who would be interested in doing something like 3 + 2, 3 + 4, and 3 + 5 all day long?

Computation question do not train problem-solving ability

Most people would think that pure computation problems do not train students' thinking ability. It is true by looking at the most workbooks purchased, but if the computation problems are designed, especially then computation problems can also train students' thinking ability and they are quite challenging. A learning centre needs to develop these kinds of math materials to keep students' interest high when working on math computation problems.

Student's name: ________________ Assignment date: ____________________

Gifted math program only works if students have mastered basics.

If students cannot get the answer 12 – 9 quickly or need to hand calculate $42 \div 2$ then they do not have number sense and mental math ability. It would be difficult for students do well if students have to use calculators to get answers for the results of 2-digit subtract 1-digit. Further difficulty when going to higher grades is students need to do even simple calculations by hands instead of by their brain, for example, to divide $4x^3 - 11x^2 + 0x + 12$ divided by $x - 2$, we have seen some students could not do $-11x^2 - 8x^2$ mentally but need to use hand calculation to actually write it out separately to just get the answer for $-11x^2 - 8x^2$.

Gifted math materials must train students' multi-task ability

One difficulty students encounter when doing math is they cannot do multiple tasks when in fact in real life many children can do multiple tasks. For example, a problem was given to student for experiment as follows: calculate $48 \div 13 \times 1 \div 24 \times 26 - 2\left(\frac{3}{4} - 1\right) + 1.5\% \times 1\frac{2}{3}$. Students need to develop the ability of handling multi-task to solve this problem.

Summary

In summary, a learning centre must develop a program which whose contents can last for 2 hours without creating boredom for students, so the material must be challenging, fun, and entertaining. The materials must be able to teach students with handle multi-level and mixed abilities. Computation problems must be designed in a way, they can also train student' thinking abilities.

Student's name: ________________ Assignment date: ____________________

Comparison between Ho Math Chess worksheets and traditional worksheets

What's wrong with the traditional drill computation? From my tutoring point of view, there is nothing wrong with giving them to children for practice on fluency and grasp of basics. However, there is something wrong with children's point of view, that is they are boring, dull, and not fun. Why do children feel that way? Well, the time has changed, but the format of traditional computation worksheets has not caught up with the pace of society. The future belongs to a generation that understands how to process information, and the information might include digits, bytes, numbers, graphics, images, languages, symbols, equations etc. Children today might chat with others on the internet while downloading or uploading files and viewing movie clips at the same time. Multi-tasking and the multi-way of processing information seem to have come as second nature to children, but is our computation format reflecting how children live today? Certainly not. This is one reason why children feel so bored and lost interest in continuing to work on the same "old" style of computation worksheets with pencil and paper.

These simple monotonic basics computation worksheets are no longer reflecting the real world the young generation is facing today or will be living in the future. The computing world children are facing today is much like a rich tapestry, where diversified fabrics and colours are integrated. Children today are absorbing not just numbers but an array of information like image, sound, music, symbols, spatial information, or even abstract ideas all bundled together and delivered through many types of media. Children today are not happy just working on pure number drills without any other stimulus or motivator. Realizing the importance of having fun while learning, Ho Math Chess has been embarked on an important teaching philosophy that is to integrate chess and puzzles into math worksheets so that children can learn math while having fun.

Ho Math Chess has created a series of special printable workbooks to have a synergetic effect by integrating or converging arithmetic basics computing, chess, mazes, and information processing all in one worksheet in an information technology environment. This is accomplished through Ho Math Chess's own trademarked proprietary technologies such as Geometry Chess Symbol, Frankho ChessDoku, Frankho ChessMaze, a brand new chess training set.

With this new invention of math, chess, and puzzles integrated worksheets, a child is acting as a data warehouse manager and sorts, matches, or classifies data through simulated cell phone screen using an incredible variety of learning concepts, namely chess, symbols, spatial relation, logic, comparison, tables, patterns, mazes, computing etc. by networking all kinds of information together, the learning process is much like to search information using the internet. Only when children have successfully followed through instructions (SCL) and, as a result, created a question themselves can a solution be found at last.

Why buy a Ho Math Chess Franchise?

Intellectual Properties of Ho Math Chess

何数棋谜 - **www.homathchess.com**

Student's name: ________________ Assignment date: ____________________

In Ho Math Chess worksheets, the questions are not written out for children but must be mined through a data warehouse, and answers must be computed by following a series of spatial relations and then analyzed using the logic pattern to reach a conclusion. The data warehouse is shaped much like a chessboard, and the chessboard is simulating a cell phone screen (See figure below). Only after children observed how data is moving through a miniature chessboard, using Ho Math Chess invented Geometry Chess Symbols, can both of the problems and answers be found.

Ho Math Chess workbook is a multi-function workbook. It trains children not only their basic computing ability but also trains them to be an astute data warehouse manager or excellent data miner by developing their problem-solving ability and critical thinking skills.

Ho Math Chess workbook provides education and also entertainment value to get young children involved in the future world they will be facing.

Student's name: ________________ Assignment date: ____________________

Image comparison of Ho Math Chess worksheet to cell phone screen

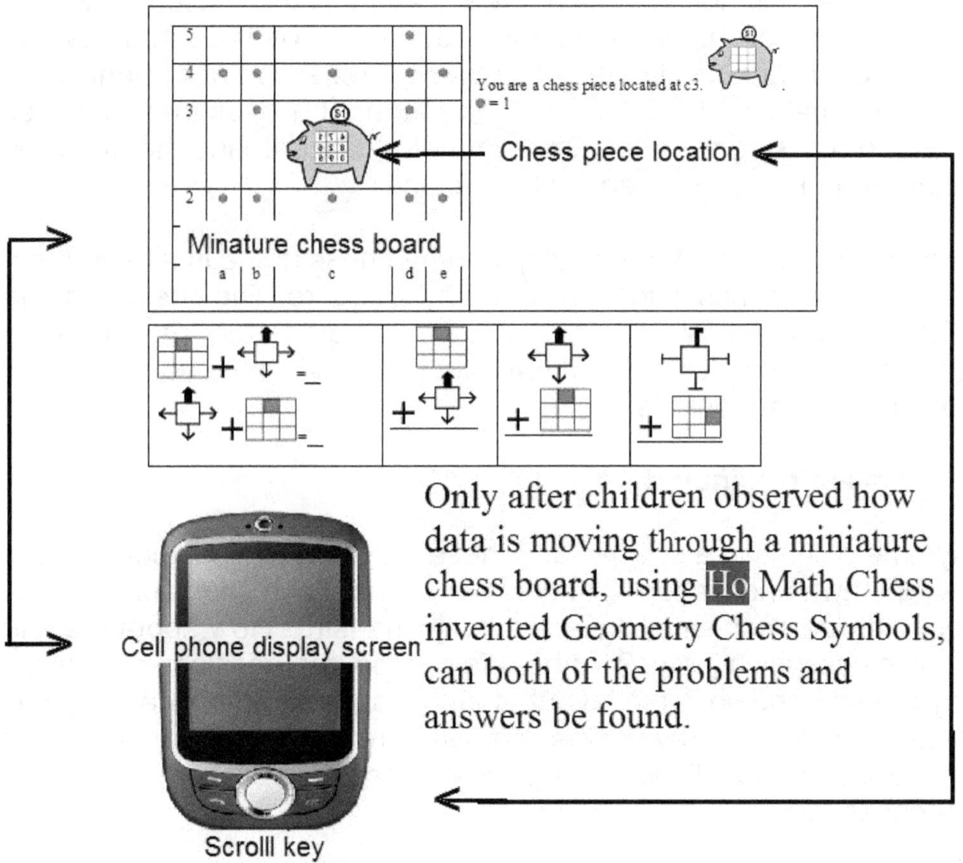

Student's name: ________________ Assignment date: ____________________

Discovery of the Key Linking Math and Chess

If one thinks that chess game is made of warriors or commanders and kings and queens battling in the field, then this notion does not really address the reason why chess pieces move in a pattern-like direction, `w6for example, the rook moves up and down or left, and right or bishop moves diagonally. Is the chess game a reflection of ancient war, or is it an invention based on a mathematical principle? The author believes that how chess is played is probably invented by using the geometry symmetry concept, and the author's conjecture is based on analyzing the moves of each chess piece.

Perhaps it is not coincidental that the playfield of the chess game is all about squares, and the Chinese character of a "rice field" is also a 2 by 2 square. The chessboard is a tessellation of 8 by 8 squares. With the same length of parameter, the square gives the largest area (battlefield) and also square forms tessellation, so the square becomes the battleground for many board games.

How chess moves are originated

There are 4 symmetry lines in a square, and these 4 symmetry lines constitute the moves of rook, queen, king, pawn and bishop. It makes sense that each chess piece moves along the symmetry lines to divide each small square evenly and fairly. How about the knight? Why knight is the only piece which jumps? To be a fair game, the positions of chess pieces must be placed in symmetry, and so is the layout of a chessboard. To play an asymmetric game, the smallest board required is 5 by 5. Chess moves are closely related to geometry translation. For example, the following is a typical description of how each chess piece moves in most chess books.

Student's name: ________________ Assignment date: ________________

Chess pieces names and moves

Symbol	Names of chess pieces	How does it move (If it is not blocked and is safe to move.)
♕(Q)	Queen (major piece)	Up and down Left and right Diagonally Any number of squares
♔(K)	King	Up and down Left and right Diagonally one square at a time
♖(R)	Rook (major piece)	Up and down Left and right Any number of squares
♘(N)	Knight (minor piece)	L-shape or Y-shaped in 8 directions The only piece can jump over pieces.
♗(B)	Bishop (minor piece)	Diagonally Any number of squares
♙	Pawn	One or two squares forward on the first move and only one square forward after the first move. One square diagonally when taking an opponent's piece. When a pawn reaches the other end of the board, the pawn can be promoted to any piece other than a king or a pawn.

I believe that the possible movements of each chess piece are originally intended to be a 360-degree circular movement. For example, take a look at a chess diagram. If a chess piece is placed at c3, how many ways can this chess piece reach out to the side of a square to form the shape of an inner circle (inscribed circle c2, b3, c4, d3) or outer circle (circumscribed circle c1, a3, c5, e3)? Depending on how points are connected, the shape of the square could also be formed.

The first "easy" way would be to move top-down or bottom-up and left-right or right-left, from

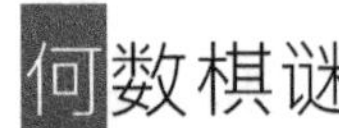

Student's name: ________________ Assignment date: ____________________

c3 so as to reach the limit of a square and an inner circle is born, and thus the movement of the rook is born. The motion of its move is called translation or slide. Connect the 4 out reached points with 4 straight lines. The shape is actually a square, but with a contour curve, then it forms a circle.

The second way of moving to the outer limit of a square and form a circle is to move in the directions of two main diagonals. Thus an outer circle is produced, and the movements of a bishop are born. Arguably, the four points also make the shape of a square. This motion from c3 to each of the 4 diagonal points is also a double-slide. The bishop can view in 360 degrees.

Combine the above two ways of rook and bishop moving. We have the most powerful move in all chess pieces that is a queen. King can only move in one square in each move and follows the moves of a queen.

In a 5 by 5 chessboard (see Figure 1), we notice that all chess squares on each of 4 sides are covered by the moves of rook and bishop except a2, a4, b1, c5, d1, d5, e2, e4, so from a game point of attacking or defending view, this is a problem – there are 8 squares which are not covered. This is the reason of the birth of another chess piece called knight which covers the 8 squares by jumping to those 8 squares because it does not move by following the same moves of rook or bishop to reach the 8 squares. This perhaps is the reason why knight jumps since knight does not trace any squares in one straight line to reach any one of those eight unreachable squares.

By using the moves of up/down, left-/right, diagonals, and diagonal jump, every square on a 5 by 5 chessboard is completely covered from c3.

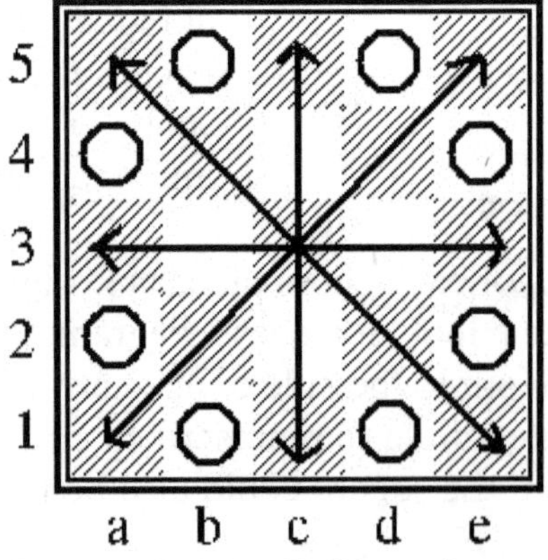

Figure 1 5 by 5 Chessboard

This geometric view of teaching chess movement in 360-degree to form a complete square is much clearer than introducing knight moves in L shape and it also explains why knight

Student's name: ________________ Assignment date: ________________

jumps – to defend or attack 8 squares where all other chess pieces are not able to achieve.

Discovery the key linking math and chess

I believe that my discovery that chess moves using symmetry property on a square is the secret key that links math and chess. For many thousands of years, chess pieces have been made of figurines, and thus, this key which links math and chess was not discovered. This actually hinders the learning of chess for very young children. The difficult task when teaching children as young as kindergartners are it understandably takes considerable more time for such young children to be familiar with how each piece should move. Without mastering chess moves, they cannot enjoy the joy of playing chess. Often it even becomes frustrating and discouraging for some children to pursue further. The main reason that children can not master the moves of each chess piece quickly is there is no clear relation between chess moves and its chess figurine. It does not seem to make sense for children why a Rook would move left-right and up-down. Based on my discovery, I created a copyrighted Geometry Chess Symbol (GCS) using the concept of geometry and also a new chess teaching set, which is created based on the Geometry Chess Symbol. The response of using GCS to link math and chess and the testing of this new chess set at Ho Math Chess are very well received. Children could start to play chess almost immediately, right after being pointed out that they just have to move each piece according to the moves marked on each piece. This incredible chess teaching set just plays like an ordinary 3-D chess set but offers an additional advantage that is the moves of each chess piece are clearly marked on its flat surface to make chess not only easier to learn but also fun for children. It is a "what you see is what you move" chess set.

The geometry concepts of lines, line segments, transformations, and intersections are used to design this revolutionary set. It is a great pattern tool to train children's skills in observation, orientation, decision, and acting with its turn point in the middle of each chess piece. Children can picture themselves at the intersection of either lines or line segments and then move according to the directions pointed by the arrows or the orientation line segments. All move directions are in line with instructions of any typical chess book. For example, the knight moves in L shape, starting with 2 long squares and then makes a turn and ends in one square move.

No more spills or bumps for small hands when moving pieces. An additional advantage with this pocket-sized, flat-surfaced set is to play Blind Chess, which uses a very simple move rule and is fun to play (more details on how to play Blind Chess, see the end of this article.). The GCS I invented, and the new chess teaching set is illustrated as follows.

Student's name: ________________ Assignment date: ____________________

Figure 2 Geometry Chess Symbol and Ho Math Chess Teaching Set

Math and Chess Integrated workbook

With this training set, learning chess has become much easier and more interesting. Why? Because these geometric chess symbols can also be incorporated into Ho Math Chess integrated workbooks, so there is a smooth transition from playing chess to working on paper and pencil math and chess integrated worksheets.

If the child is not ready to work on math and chess integrated workbooks, then this still is a great chess set to facilitate learning on how to play chess. The following example demonstrates how this Geometry Chess

Why buy a Ho Math Chess Franchise?

Intellectual Properties of Ho Math Chess

何数棋谜 - www.homathchess.com

Student's name: ________________ Assignment date: ________________

Symbol is used in the math and chess integrated workbook. This sample demonstrates how the Geometric Chess Symbol (GCL) is integrated with math after children used the Ho Math Chess Teaching set; it also shows how subtraction facts are related to addition. Further, the commutative property is taught to children. Multi-direction computing is learned here. No questions are given for children but require them to go through orientation and "create" the problem themselves. This reinforces the learning instead of drilling but has the effect of "drilling", but not boring.

Addition and subtraction

c	Ѱ	2	4
b	3	Ѱ	6
a	7	8	✥
	1	**2**	**3**

Move one square at a time from the original square b2, which is represented by the following symbol: □.

For the following problem #1 only,
□ = Ѱ = 1, circle = 2 and the triangle =3.

✥	□ + ○ = △ △ - ○ = □ + ○ + □ = △ △ - □ = ○ □ + ○ = △ = ○ + □ = △
✥	□ + ○ = △ △ - ○ = □ + ○ + □ = △ △ - □ = ○ □ + ○ = △ = ○ + □ = △
✥	□ + ○ = △ △ - ○ = □ + ○ + □ = △ △ - □ = ○ □ + ○ = △ = ○ + □ = △

The invention of the math and chess training set has revolutionized the chess learning population profile to as young as four years old, and children can learn to play in less than an hour.

Student's name: ________________ Assignment date: ____________________

Student's name: ________________ Assignment date: ___________________

Do Chess and Math Have Anything in Common?

Chess has many math characteristics that are mostly learned in elementary schools. There are other math-related topics such as search tree of possible moves etc., which are high-level thinking math and is related to artificial intelligence. Definitely, chess and math are related in various math subjects and areas. My research area is in elementary mathematics and chess, so I would like to talk about how chess and elementary math have something in common and perhaps also touch on topics on the areas that are unique in their own way. In other words, they are totally not in common.

Chess deals with the whole number system, so it is very comfortable for a young child to learn chess and at the same time to learn our whole numbering system. The chessboard is normally square in shape, and players use chess notation, this definitely is "borrowing" the concept of coordinates from math, but it only serves to benefit the child. Image a 5-year-old is learning the X and Y coordinate system math concept by learning chess?

The concept of relative values of chess pieces is similar to the concept of using unknowns in algebra, and yet the chess values are very meaningful to children. Why could a queen be numerically substituted by a value of 9? This constant value is the most beautiful concept that we could use to teach a child to learn the concept of substitution and function while they are having fun.

Chess provides a conduit to lead a child to learn some math concepts without any hard pressure. They practice their critical thinking skill while pondering the next best move, considering all the pros and cons; weighing all the possible moves. This process of thinking involves data gathering, analyzing, synthesizing and integrating, which are all critical thinking skills.

The benefits of using chess as a means of developing critical thinking while other teaching "tools" are not readily available is a child can constantly get feedback from the opponent while they socialize and entertain each other. The wrong moves could get penalized by the opponent –immediate feedback of their critical thinking skill. No other critical thinking training course could provide such an instant response and so much entertainment value.

The checkmate and check pattern is just amazing, and there are no other substitutes that could be used to train youngsters in this kind of pattern recognition and spatial relation. A number of 3 added to another number of 5 is just plain 8 in math workbooks. In chess, this calculation is not just a vertical 3 plus 5 or horizontal 3 plus 5 we see most of the time in worksheets. It could involve a rook and knight attacking the same piece, and the directions are multi-direction and the way to solve this type of question involves spatial relation and pattern visualization.

Student's name: _______________ Assignment date: _________________

The chess pattern and math pattern is so different that I have not been able to find anything in the math pattern which could replace the "cause and effect" pattern in chess since the chess position has a purpose that is to check or checkmate or attack, defend etc., but this kind of pattern does not exist in math.

The distance of some chess moves does not make sense at all from the math point of view. How could a knight take more moves to reach a closer square than a faraway square? This is intriguing. How could a pawn reach the last rank takes the same number of moves to reach it from its diagonal? Isn't the slant line longer than a straight line? This does not make sense at all. But it all works out beautifully in chess, and it even becomes a famous counting square problem in chess.

Chess players are constantly looking for the mate position by coordinating chess pieces in cross lines, isn't what a student is trying to find for the solution of system equations in math? This is also the concept of set theory.

How many symmetry lines can a square have? This is the answer to how a queen moves. Does a knight really move in L shape? What happens when the knight is looking for the next move? It is really looking out directions almost like a circle. Chess really is not a game of rank and file; it is a game of circle (movements) and square (chessboard).

Many interesting mathematical problems could be created if one truly appreciates the beauty of the math concept built-in chess. The math and chess integrated problems not only advances a child's chess knowledge; it also improves a child's ability in problem-solving, critical thinking, logic, and visualization.

Ho Math Chess Learning Center has created the world's first commercially available math, and chess integrated workbooks, and more details can be found by visiting www.mathandchess.com.

Student's name: ________________ Assignment date: ____________________

El Ajedrez beneficia a los niños

¿Por qué les fascina el ajedrez a los niños? Dr. Montessori observó que los niños pequeños se ven muy atraídos por los instrumentos de desarrollo sensorial. Dado que el ajedrez es multisensorial y práctico, involucra coordinación ojo-mano-cerebro en múltiples direcciones que no son lineales, si las comparamos con la mayoría de los video juegos o los juegos del computador. Cuando mi hijo tenía 5 años, conduje a mi hijo hacia que desarrolle la capacidad de capturar mis piezas de ajedrez como su recompensa: quizás esta experiencia sensorial fue lo que le produjo sed por más juegos.

Además de la diversión, al jugar ajedrez se logra el desarrollo de destrezas cognitivas y de pensamiento crítico, habilidades de razonamiento y solución de problemas, focalización, visualización, destrezas de análisis y planificación. La investigación reciente corrobora estas conclusiones (1)

¿En qué se diferencian mis rompecabezas de otros?

En el pasado, se han publicado muchos rompecabezas. Sam Loyd, el "Rompecabezas Rey," compuso los problemas de ajedrez más paradójicos y fenomenales. El libro más reciente de matemáticas y ajedrez titulado *Mathematics and Chess* tiene 110 problemas entretenidos con sus soluciones. (2) Casi todos estos rompecabezas publicados, se relacionan con movidas de las piezas de ajedrez y la mayoría se consideran muy difíciles para estudiantes de primaria.

Una serie de libros de matemáticas llamado *Challenging Mathematics* (3) incluye al ajedrez como parte de una sección de lógica pero el contenido de ajedrez en si está separado y no integrado con los conceptos matemáticos o los problemas.

Para crear mis rompecabezas de ajedrez, usé símbolos de ajedrez, valores de ajedrez, movidas, tablero, notaciones algebraicas, ubicación de las piezas, conteo de ataque y defensa, orden de intercambios, y reglas de ajedrez. La diferencia fundamental entre los rompecabezas de matemáticas y ajedrez, y los publicados tradicionalmente, es que los símbolos de ajedrez, las movidas y los valores están integrados a las matemáticas para crear problemas sobre patrones, lógica, geometría, rutas de conteo, relaciones, arreglos, numeraciones y hasta manejo de datos. De esta forma, los niños de prekinder hasta primaria, tienen, mientras aprender a jugar ajedrez, una oportunidad para explorar rompecabezas de matemáticas haciendo uso de su conocimiento básico de ajedrez.

Los rompecabezas están diseñados para mejorar las habilidades matemáticas usando el ajedrez como herramienta de enseñanza. No pretenden reemplazar la enseñanza escolar de matemáticas sino que sirva como un material complementario, para enriquecimiento. Los niños aprenden mejor cuando juegan, con de juegos de mesa. Math + Chess = una forma divertida de aprender matemáticas.

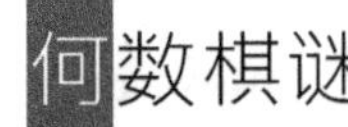

Student's name: ________________ Assignment date: ____________________

Cómo integramos las matemáticas y el ajedrez

La creación de problemas de matemáticas y ajedrez demanda un claro y profundo dominio del ajedrez y del currículo de matemáticas escolar, según grados escolares. Solo con este conocimiento, además de una mente creativa, se puede crear rompecabezas de matemáticas y ajedrez con significado. Los problemas de matemáticas y ajedrez están integrados usando los siguientes principios:

(1) El tablero y las piezas de ajedrez (Ver figura 1.)

El tablero es simétrico en sus diagonales centrales en términos de su color. El tablero de ajedrez es hecho de cuatro tableros pequeños idénticos, si lo dividimos por una línea horizontal y una vertical, hacia el centro. La posición inicial de las piezas de ajedrez es simétrica entre el Negro y el Blanco. La posición de las piezas de ajedrez en cualquiera de los lados es palíndroma, excluyendo el Rey y la Reina.

Los rangos y las líneas están relacionados a coordenadas. Cuando una pieza es atacada y defendida, requiere algunos cálculos aritméticos en términos de un número de piezas de ataque o defensa. Esta es la primera lección que el niño aprenderá sobre "contar".

(2) Movidas del ajedrez

Una movida de la torre es, en geometría, un movimiento en deslizamiento o traslación (izquierda/derecha, arriba/abajo) (izquierda/derecha, arriba/abajo). Las movidas de la Torre, en el ínterin, antes de llegar a su destino son similares al concepto conmutativo. Por ejemplo, la movida de a1 a h1(7) = a1 a c1 (2) + d1 a h1 (5) = h1 a d1 (5) + c1 a a1 (2).

Para dibujar la "mejor" movida, uno necesita encontrar todas las posibles rutas. Un concepto similar en matemáticas será el uso de factor 3 para hallar los factores primos de un número.

¿Cómo hacemos jaquemate a nuestro opositor? Si la Torre está en a1, y está libre para moverse a lo largo de la fila a y rango 1, qué tenemos que considerar antes de movernos? Ver si cualquiera de las piezas del oponente que interseca a la Torre, sería como hallar el valor de y cuando $x = 1$. Las posiciones de jaquemate son en realidad las intersecciones de los rangos o filas, las cuales son muy similares al concepto de solución de ecuaciones por el método gráfico.

El hallar los cuadrados comunes (cuadrados que ambas piezas pueden controlar) es similar a la idea de hallar factores comunes de dos números.

Student's name: ________________ Assignment date: ____________________

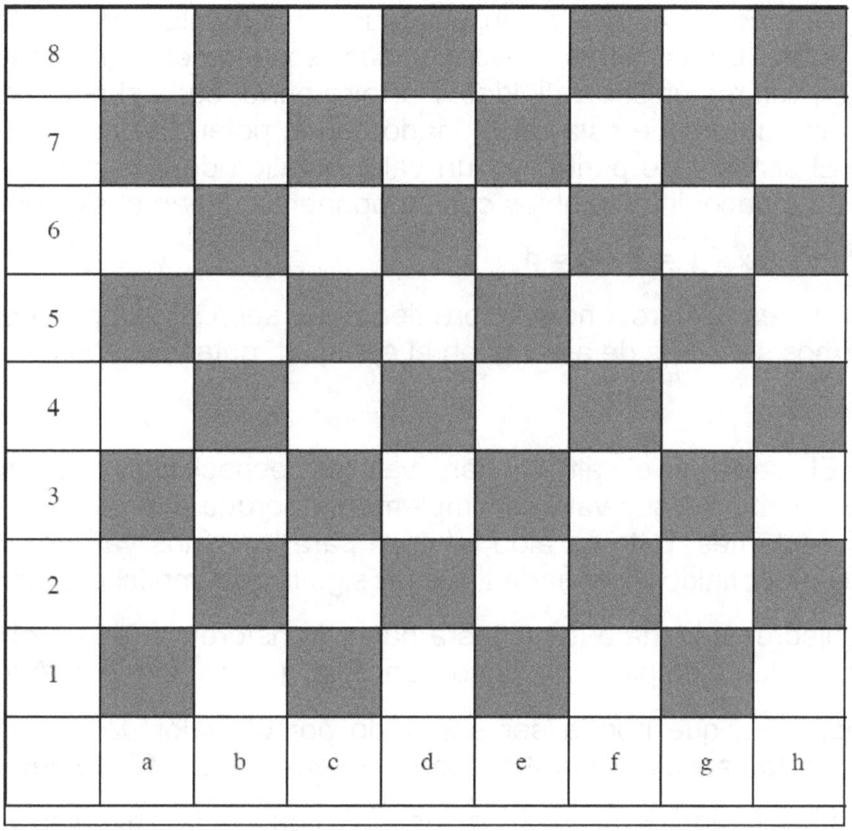

8								
7								
6								
5								
4								
3								
2								
1								
	a	b	c	d	e	f	g	h

Figure 1

Uno podría pensar que el ajedrez no tiene nada que ver con las fracciones, ya que las movidas están todas en números enteros. Por qué la reina es la pieza más ponderosa y por qué tendemos a mover las piezas hacia el centro? Todo esto tiene que ver con la relación *a*/64 donde *a* es el número de cuadrados bajo control

Cuando los jugadores de ajedrez ven las posibles movidas, la vista viene a ser un movimiento circular en 360°. Por ejemplo, al revisar las posibles movidas de la torre, un jugador escanea los siguientes ángulos: 0°, 90°, 180°, 270°. En otras palabras, la movida de la Torre, es equivalente a rotar la Torre en cuatro direcciones. El mismo concepto de rotación se aplica para el Alfil, la Reina y el Rey.

(3) Símbolos de Ajedrez y Valores de Ajedrez

El uso de Símbolos de ajedrez as Constantes

Why buy a Ho Math Chess Franchise?

Student's name: ________________ Assignment date: ____________________

Las letras romanas tales como *x, y, y z* son usadas normalmente para representar valores numéricos desconocidos. Estas letras desconocidas son también llamadas variables y normalmente no tienen valores únicos definidos. Por otra parte, cada símbolo de ajedrez tiene un significativo valor en puntos que está relacionado con la potencia que tiene cada pieza en el juego de ajedrez. El sistema de puntos es un valor estático de una pieza y generalmente sirve como una guía para hacer intercambios con su oponente. Vean el siguiente ejemplo:

Si $x = 1$, y $y = 3$. Entonces $x + y = 1 + 3 = 4$.

En el ejemplo, *x es* 1 y *y es* 3. Pero *x* no siempre tiene que ser 1 ni *y* tiene que siempre ser 3. Son variables. Si usamos símbolos de ajedrez en el ejemplo, obtendremos:

♙ + ♗ = 4

El peón y el alfil tienen valores específicos definidos en 1 y 3 respectivamente. No cambiarán sus valores simplemente porque el problema es diferente: en otras palabras, son constantes. Esto es algo intuitivo para los niños ya que el valor de cada símbolo de ajedrez es predefinido y por ende tiene un significado implícito a ellos.

Uso de símbolos de ajedrez enseña a los niños a hacer transferencia de un objeto concreto a un concepto abstracto. Por ejemplo, un objeto concreto, (como la pieza Alfil), puede estar asociado a un símbolo ♗, que podría ser sustituido por un valor de 3. Este proceso de aprendizaje y pensamiento está alineado al concepto de enseñanza de Montesori.

Cuando comparamos los símbolos de ajedrez, usando figures de animales o cualquier otro símbolo tales como *x*, *y y z en rompecabezas de mate y ajedrez,* tendrá menos significado para los niños. Los niños no se confunden con los símbolos de ajedrez, ni se volverán inútiles cuando aprendan las variables en la secundaria, simplemente porque aprendieron sustitución en edad más temprana. Los símbolos de ajedrez son usados solo como pictogramas o representaciones.

La otra razón para usar símbolos de ajedrez en rompecabezas de matemáticas y ajedrez es que los símbolos de ajedrez representan movimientos. Coincidentemente, algunas direcciones de movimientos se asemeja a algunos operadores aritméticos: por ejemplo una Torre se puede mover arriba, abajo, o izquierda, derecha, y así su trazo de movidas posibles se ven como un signo más + .

Cada símbolo de ajedrez tiene una dirección de movimiento especialmente definida y estas direcciones son interiorizadas en cada pieza. He aprovechado estas movidas de las piezas de ajedrez' y las he definido de la siguiente manera:

Suma/Resta = Torre (También puede ser Reina o Rey)
Multiplicación = Alfil (También puede ser Reina o Rey)
División = Rey (Oposición de dos Reyes)

Student's name: ________________ Assignment date: ___________________

Valores de Ajedrez Usados

Los valores de símbolos de ajedrez son los mismos a los usados en el Manual de *Chess Teaching Manual* -Federación de Canadá. (4)

La técnica de eliminación para contra los puntos de piezas de ajedrez serían muy similares al concepto de la propiedad de resta de una ecuación.

Los símbolos de ajedrez y los valores son integrados a operaciones aritméticas para crear un nuevo tipo de problemas. Mi propósito al usar símbolos de ajedrez es crear un tipo de preguntas más interesantes y motivar a los niños a que piensen mejor, y a que usen razonamiento espacio temporal y solución de problemas.

Cada pieza de ajedrez tiene asignado un valor con puntaje, tal como se muestra en la siguiente tabla.

♔ (king) = 0 point	♘ (knight) = 3 points	♖ (rook) = 5 points
♙ (pawn) = 1 point	♗ (bishop) = 3 points	♕ (queen) = 9 points

Mi experiencia en usar valores de ajedrez para enseñar operaciones aritméticas ha sido muy positiva. Los estudiantes de primaria que no han aprendido variables, pero que han trabajado en mis hojas usando símbolos de ajedrez han absorbido el concepto de variables algebraicas o sustitución en una forma natural e intuitiva. No hay necesidad para explicar el concepto de variable sino solo mencionar los valores de las piezas de ajedrez. Considere el siguiente ejemplo.

♖ + 5 = ______

Student's name: ________________ Assignment date: ____________________

Ejemplos

A continuación tenemos unos ejemplos creados para ver la relación entre matemáticas y ajedrez

□	□	□	□
	− ♖		− ♖
______	______	______	______
+ ♗	♕	+ ♖	6
♕		6	

Student's name: ________________ Assignment date: ____________________

Example 2. Multiplication

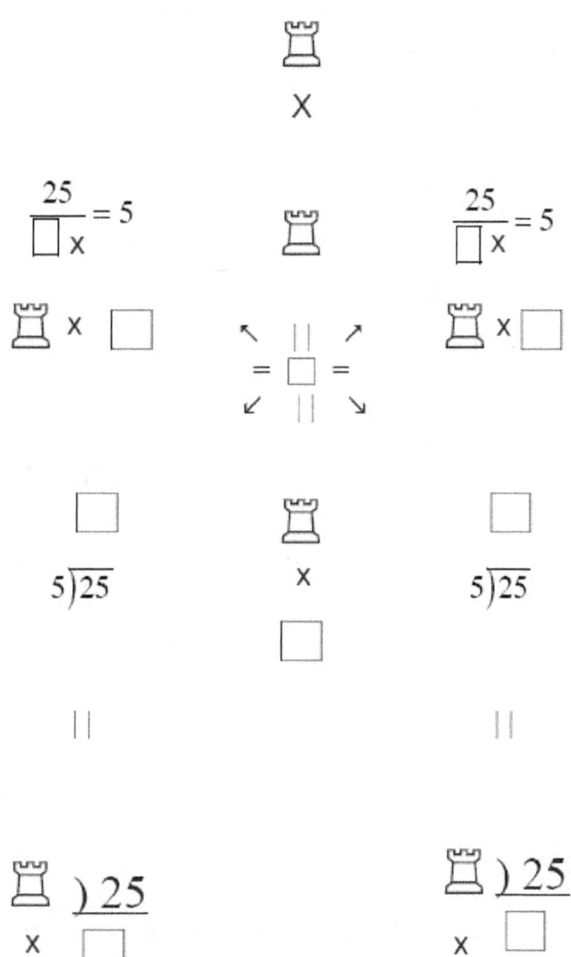

I created the above problem with the view that in real life, children do not really lean subtraction, multiplication, or division sequentially. So here I incorporate the idea of mul using different formats of computing. The purpose of this worksheet is not only to learn mu but also to expose children to how multiplication could be written in different ways.

Student's name: ________________ Assignment date: ____________________

Example 3. Logic

<table>
<tr><th colspan="2">Chess Symbol</th><th>Logic Training</th></tr>
<tr><td colspan="2">New chess symbols are defined as follows.</td><td rowspan="8">In the following equation, observe the chess symbols on the left and fill in each ◯ with a number.

If ┼ + ┼ = 10

then ÷ + ┼ = ◯

   ┼ ✳
+ ┼ ÷
―――――
◯ ÷ ✳</td></tr>
<tr><td>Chess figurines</td><td>Chess symbols</td></tr>
<tr><td>♔</td><td>÷ (Opposition)</td></tr>
<tr><td>♖</td><td>┼</td></tr>
<tr><td>♘</td><td>L</td></tr>
<tr><td>♗</td><td>×</td></tr>
<tr><td>♕</td><td>✳</td></tr>
<tr><td>♙</td><td>⩛</td></tr>
</table>

The above problem, suitable for Grade 3 and above, demonstrates to me that children can be led to correctly solve problems using additional "creative" chess symbols.

Using the above Chess Symbol table, find the following pattern. Z, ÷, O, ⩛, T, L, T, ×, F, ┼, ______ , ✳
Using the above Chess Symbol table, find the following pattern. 0, ÷, 1, ⩛, ____, L, 3, ×, 5, ┼, _____, ✳

Why buy a Ho Math Chess Franchise?

Intellectual Properties of Ho Math Chess

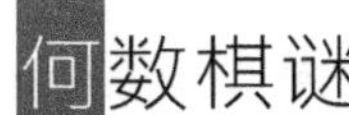
何数棋谜 - www.homathchess.com

Student's name: ________________ Assignment date: ___________________

La mayoría de los niños no pueden resolver el rompecabezas. Sin embargo, son conscientes de la sofisticación de los problemas cuando les explico las relaciones lógicas. Estos rompecabezas usan una combinación de símbolos de ajedrez y sus valores.

El uso de valores de ajedrez es muy parecido al uso de valores monetarios. Cuando los niños los ven, las figuras de ajedrez o de monedas representan valores predefinidos y con significado. El siguiente es un ejemplo donde los valores de las piezas de ajedrez puedan ser valores monetarios y el "Puntaje Total" pueda ser una suma de valores.

Ejemplo 4 Tabla Valores

Llene el número de piezas de ajedrez para obtener cada total.

Number of ♙	Number of ♘	Number of ♖	Total points
1	1	1	9
3	2	0	9
0	3	0	9
☐	☐	☐	10
☐	☐	☐	10
☐	☐	☐	11
☐	☐	☐	12
☐	☐	☐	13
☐	☐	☐	14
☐	☐	☐	15

Ejemplo 5. Ecuaciones

Los siguientes ejemplos demuestran cómo los símbolos de ajedrez y valores de ajedrez son integrados con operaciones aritméticas.

$$♕ + ♘ + x = 54$$

$$x = \_\_\_\_$$

Student's name: ________________ Assignment date: ____________________

Ejemplo 6 Suma y Resta, Si, Entonces - ....

10	−	♔	=	□		
5	+	♔	=	□		+
□				□		

If 10 + ♖ = □, then 9 + ♖ must be □

If ♖ + 10 = □, then ♖ + 9 must be □

Example 7 Cross Multiplication

	♙ =		8 =	
	□		□	
	X	⤡⤢ X	X	
	□		□	
□		+		□ = 6

Student's name: ________________ Assignment date: ____________________

Ejemplo 8 Multiplicación y división

♕ × 2 ——— 18 ÷ 2 = □	♕ × ♖ ——— □ ÷ 5 = □
♕ × ♖ ——— □ ÷ 9 = □	♕ × ♕ ——— □ ÷ ♕ = □
♕ × ♘ ——— □ ÷ ♘ = □	♕ × ♘ ——— □ ÷ ♕ = □
♖ × ♘ ——— □ ÷ ♘ = □	♖ × 8 ——— □ ÷ ♖ = □

Student's name: ________________ Assignment date: ____________________

The above operations should not confuse children. The expression ♕ × ♘ will not make sense if it is explained literally as a Queen times a Knight. However if it is translated into numerals, children will understand that they are computing 9 x 5.

Ejemplo 9. El siguiente rompecabezas requiere conocimiento de movidas de ajedrez

Filling in □ by a chess piece	Geometric shapes
♙	
♖	
□	
♗	

Student's name: ________________ Assignment date: ___________________

Ejemplo 10 Use movidas de ajedrez para resolver el siguiente rompecabezas.

En primera instancia muchos estudiantes no pueden resolver este rompecabezas. ¿Por que? Porque están tan acostumbrados a usar la operación de izquierda a derecha, pero esta pregunta debe ser usada en dirección no convencional.

Ejemplo 11 Use movidas de los símbolos de ajedrez para resolver lo siguiente.

If 2 ♖ 3 = 5 then 2 ♗ 3 is = _____

Sorprendentemente, algunos estudiantes no tienen problema en resolver el rompecabezas anterior.

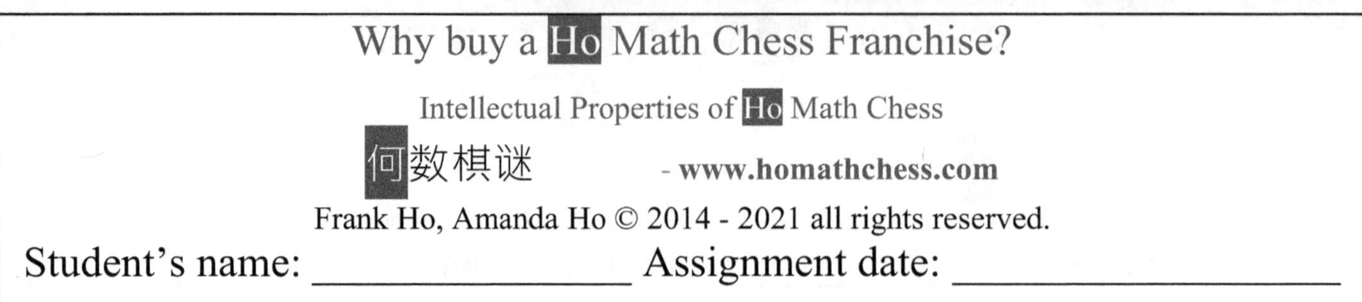

Student's name: ______________ Assignment date: ________________

Ejemplo 12

Use movidas de los símbolos de ajedrez para resolver el siguiente rompecabezas.

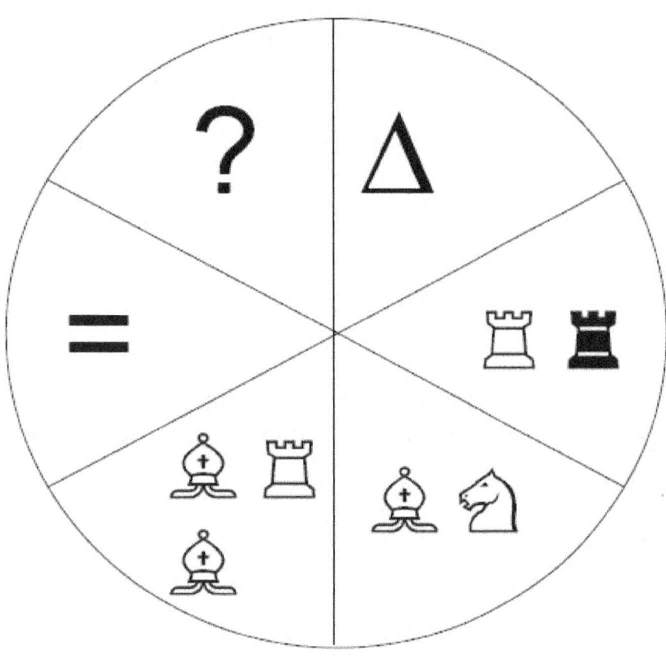

Example 13

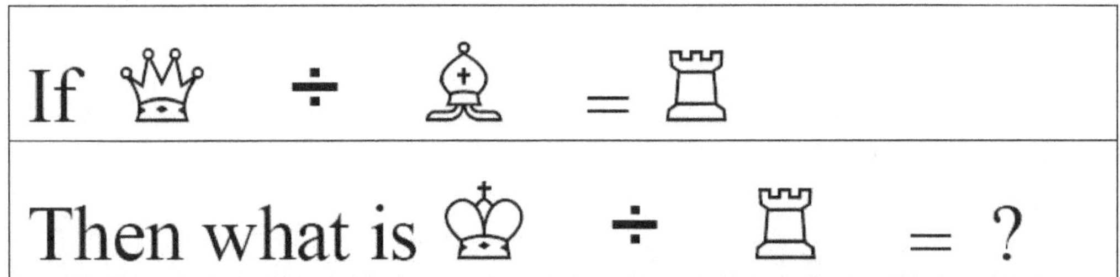

El problema no puede resolverse usando valores de ajedrez; los estudiantes tratan y lo saben hacer. Entonces cuál es el truco que está detrás de los rompecabezas?

Why buy a Ho Math Chess Franchise?

Intellectual Properties of Ho Math Chess

何数棋谜 - **www.homathchess.com**

Student's name: ________________ Assignment date: ___________________

Ejemplo 14

Ruta de la Torre

Marque con una cruz (X) el cuadrado (s) común (s) que pueden compartir todas las torres.

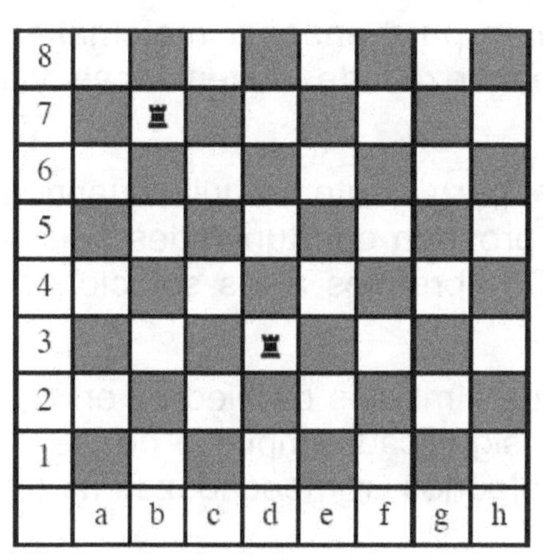

Halle el factor común de los siguientes números.

12, 24

13, 26

Marque con una cruz (X) los cuadrados comunes que pueden compartir todas las torres.

8								
7								
6								
5								
4								
3								
2		♖						♖
1								
	a	b	c	d	e	f	g	h

Halle los factores comunes de los siguientes números.

11, 121

3, 26

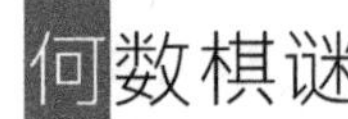

Student's name: ________________ Assignment date: ____________________

Los problemas de movidas de ajedrez pueden ser comparables con los conceptos matemáticos. Di los ejemplos anteriores para demostrar la idea con problemas de ajedrez por un lado y los problemas de matemáticas por el otro.

Ejemplo 15

Halle los valores y reemplace ? o llene en el cuadrado vacío.

Resumen

He comprobado que la idea de usar símbolos de ajedrez para enseñar matemáticas a los niños es muy útil: pueden aprender a resolver rompecabezas de matemáticas y ajedrez usando su conocimiento de ajedrez.

Muchos no pueden jugar ajedrez bien contra su oponente, pero sienten orgullo cuando pueden resolver los rompecabezas de Math and chess ya que proveen oportunidades adicionales a algunos de los niños para que enfrenten a desafíos. Doy premios a los solucionadotes de rompecabezas ganadores.

Lo más importante de usar símbolos de ajedrez es que los símbolos de ajedrez en si, no solo poseen valores predefinidos sino que también tienen un significado implícito de las movidas. Estas dos características especiales me permiten crear algunos rompecabezas matemáticos con pizzazz.

Usando símbolos de ajedrez, un problema aritmético de un paso puede convertirse en un problema de múltiples pasos. Como un resultado, los símbolos de ajedrez y valores ofrecen a los niños más oportunidades para trabajar en otros tipos de preguntas, pueden servir para estimular las mentes de los niños e impulsar las habilidades cognitivas de los niños. Es así que la ventaja de trabajar con este tipo de problemas es doble: mejorar el conocimiento de ajedrez y también las habilidades de solución de problemas matemáticos.

Mis rompecabezas de matemáticas y ajedrez no solo tienen que ver con la sustitución mecánica de los números por símbolos de ajedrez. Muchos rompecabezas también tienen que ver con patrones, secuencias, geometría, teoría de conjuntos, y con lógica. En otras palabras, la integración es muy diversificada y también tiene que ver con visualización en multidirección.

Para acabar este artículo les presento a continuación un rompecabezas de patrones, demostrando cómo los símbolos de ajedrez y valores son presentados en múltiples direcciones y enfoque multisensorial.

Why buy a Ho Math Chess Franchise?

Intellectual Properties of Ho Math Chess

何数棋谜 - www.homathchess.com

Student's name: ________________ Assignment date: ____________________

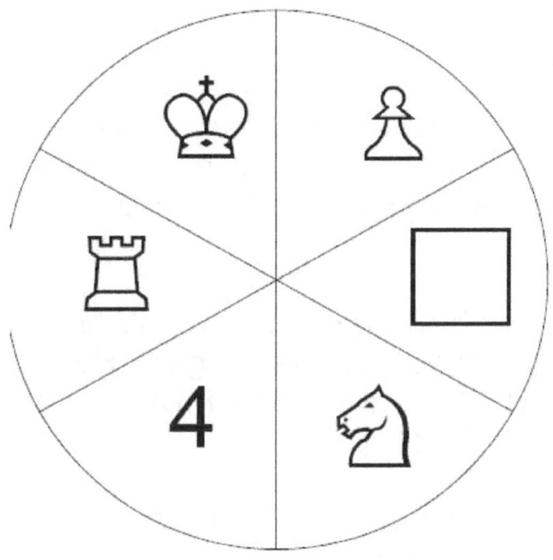

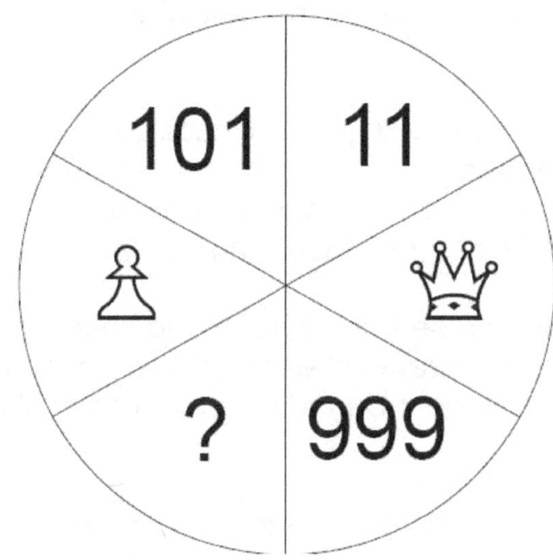

Student's name: ________________ Assignment date: ____________________

Finding Math and Chess Learning Center as Easy as 1 2 3

As the founder of Ho Math Chess Learning Center, the world's first and largest math and chess truly integrated learning center, I can speak from my own experience on how to find math and chess truly integrated learning center and why it is advantageous to learn math from math and chess truly integrated learning center.

I have been researching math and chess connection subject and had created the world's first math and integrated chess workbook because my son was interested in chess at age 6. He was the youngest Canadian junior chess champion at age 12. Later he became a FIDE chess master.

I coined the term "math and chess learning center" in 1995. Since then, the concept of integrating chess into math has become commercially popular, but those places claiming to offer math and chess classes still teach math and chess as two separate subjects, and there is no real connection between math and chess. The math and chess true integration are only possible at Ho Math Chess because I have invented the technology to make the connection happening. This is why Ho Math Chess is only the learning center where math and chess are genuinely integrated.

Why integrate math and chess? The reason is by using Ho Math Chess invented technology, we have created innovative math worksheets that are more fun for children to work with, and children are more focused and feel less boring. When working on math and chess integrated workbooks, not only can children raise their math marks, they also improve their critical thinking skills, and the most important is these workbooks are fun to work with.

How can we tell if a learning center or course labelled as "math and chess" does truly integrate chess it math? Only Ho Math Chess has the patented technology to integrate math and chess, so my honest advice is to follow the following 3 steps:

1. Investigate
   To find out if math and chess, truly integrated material, is offered.

2. Compare
   Ho Math Chess got involved in math and chess integrated teaching because its founder Frank Ho has researched math and chess teaching for over 10 years and is the authority in the field.

Student's name: ________________ Assignment date: ____________________

3. Decide
   Only after one has done a comparison and investigation can one make the correct decision.

For more details about fun math, chess and puzzles and over 100 testimonials, visit www.mathandchess.com.

Student's name: ________________ Assignment date: ____________________

何數棋谜幼兒專用棋

何數棋谜 (Ho Math Chess™) 發明了世界首創有版权的幼兒專用國際象棋.每一棋子的走法都刻在棋子上. 幼兒不但不會再為每一棋子的走法而困惑,而且還學會了幾何圖形的概念. 此一科研產品還可以讓兒童玩暗棋,不亦樂乎. 學前幼兒再也不必擔心碰撞倒棋子,而可以輕鬆快樂的學棋並享受學國際象棋的好處.

詳細情形請訪問 www.mathandchess.com.

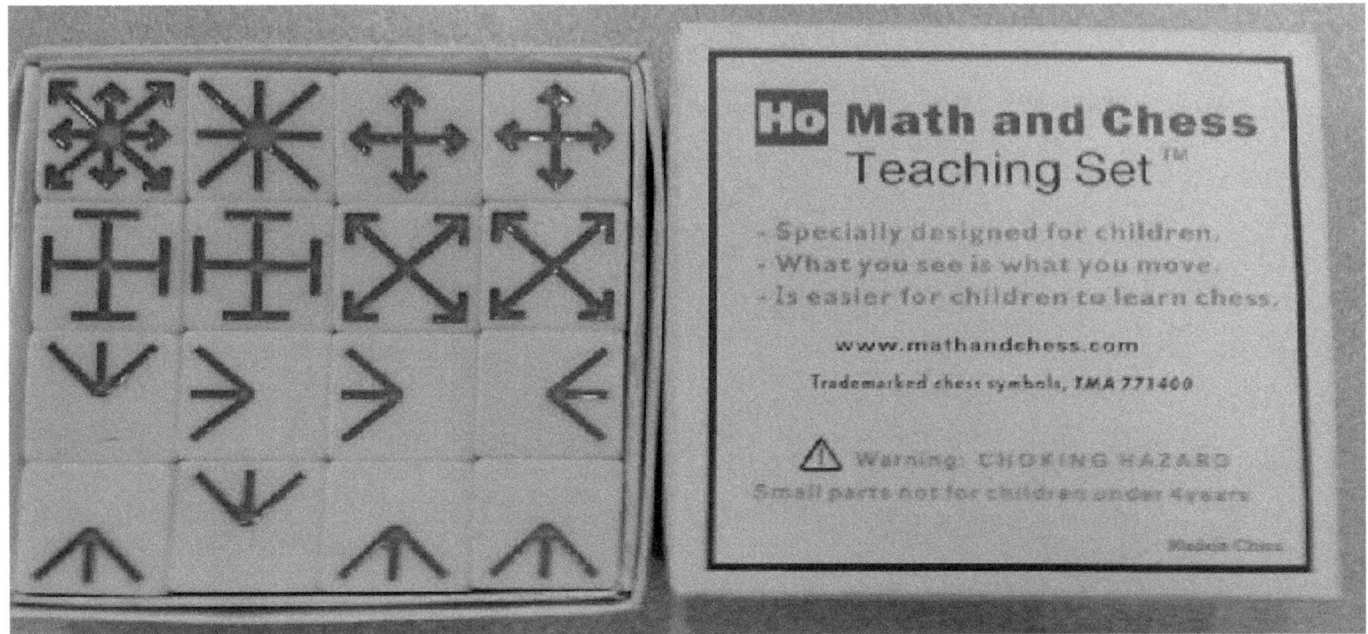

Student's name: ________________ Assignment date: ___________________

使用何数棋谜教材的好处

今天儿童面对的世界是学习如何处理数字,图形,资料搜寻,音影上下载,资讯比较,分类等资讯.这些活动实际已成為儿童生活的一部份.所以如果说学数学就是计算数字就错了.学数学的另一个目的就是学习如何数字资讯去解决问题及培养创造力.但是传统式数学的计算练习题却完全没跟上科研已经改变了儿童面对的世界.

儿童想要的计算题已经不是单纯的从上到下,从左到右的纯计算.儿童需要的是他们情愿的而又快乐地做不枯燥的计算题.所以如何将传统式数学计算题变得有趣而且又好玩,并且还可以增强儿童的计算及解决问题能力及培养创造力,同时还可以增进儿童记忆的能力达到全脑开发的目的?

何数棋謎首创巳申情商标的几何棋艺符号并利用此符号发明了世界第一的何数棋迷教材及教学棋具. 何数棋謎教材让儿童能利用几何棋艺符号进行数学的运算.

何数棋謎与传统式、数学教材不同的是小朋友不但要发掘题目,而且
还要依国际象棋棋子的走法去发掘新的方向(見下图) 及答案.

何数棋謎是将国际象棋融入数学以达到寓教於乐的教学理念.学生不但可以增强计算能力并且還可以增强解题能力及培养全脑开发创造力.

详细资料请上网 www.mathandchess.com.

Why buy a Ho Math Chess Franchise?

Intellectual Properties of Ho Math Chess

何数棋谜 - **www.homathchess.com**

Student's name: ________________ Assignment date: ____________________

Image comparison of Ho Math Chess worksheet to cell phone screen

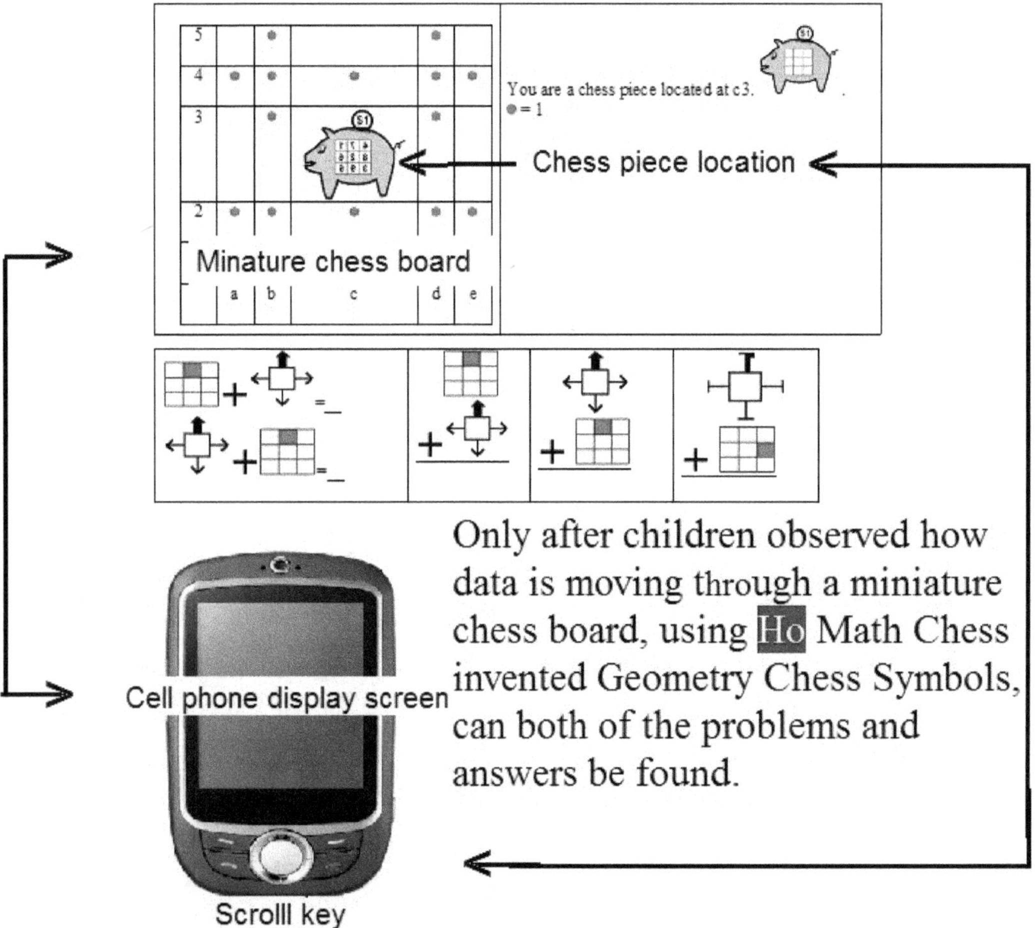

Student's name: ________________ Assignment date: ____________________

何數棋謎™: 獨特的遊戲數學教学

您是否还依稀记得儿时的兴趣所在呢？游戏，抑或是每一位在孩儿提时，都喜欢的玩耍方式。那么，利用这种游戏的方式学习数学，您说孩子是否会对其感兴趣呢？答案是肯定的。

怎么才能把游戏这种有趣味的项目结合到学习数学当中呢？用什么方式结合呢？下面笔者给您介绍一位研究此方面的专家——何凤台先生 (Frank Ho), 加拿大何數棋谜™: ™ 培訓中心創辦人。

何凤台，祖籍江苏人，从小在台湾长大，并畢業于台湾成功大学，尔后赴美留学，先后在美国 BYU 及猶他州立大学留学，并以优异的成绩获得电脑学士学位及统计学硕士学位,畢業後分別在猶他州立大学及加拿大 UBC 大學任統計電腦顧間.

數學棋藝教學理念及方法為有 BC 數学證書的何老師於 1995 年首創發明.

说起这项伟大的发明创造，还源于何老師的小儿子.

.何老師的兒子於 5 歲多時對國際象棋發生興趣,於是何老師展開了一連串的父教子的自學活動. 兒子於 12 歲時得了加拿大 12 歲組,
少年組(14 歲),青年組(18 歲)的三冠王,破了加拿大棋界的紀錄, 並且於巴西的分齡世界國際象棋大賽時得第 5 名. 兒子為 FIDE 國際象棋大師.
看到 5 歲的儿子深深的迷上国际象棋，何老師突发灵感，于是利用擅长的电脑技术，把想法负诸于实践，成功的在数学游戏教育领域取得了突破性的进展。何老師融合數學與國際象棋的教學理論巳在 BC 省數學教師刊物上發表,何老師發明的

Student's name: ________________ Assignment date: ____________________

Symbolic Chess Language 專門用來增強數學計算能力並提高兒童解題能力也申請了版权及商标.

何數棋谜教學的效果已經過統計的分析而証明對增進數學能力有顯著的效用.並同時也可以增進兒童的創造力,邏輯推理力,判斷力,自信心及上課的表達能力等多重效果.

Student's name: ________________ Assignment date: ___________________

何數棋谜自 1995 年創辦以來,已成為加拿大溫哥華最有信譽及最大的數學專科補習中心

何老師把国际象棋走法转换成一种语言数字符号，从而把枯燥无味的数学计算转化为耐人寻味的数学谜题並可以做數字的運算。因此何老師经过潜心的研究使棋艺与数学结下了深深的不解之缘！棋艺数学教材从开发的第一本教材到最后一系列教材及棋具的研發都完完整整的向政府登记申请了专利权及著作权。科学、独特、创新是此类教材的优秀所在；材料灵活，因材施教是此项创新产品的魅力所在。

教材的优越性在一定程度上提高了学生的学习兴趣。经过多方面了解，温哥华何數棋谜中心已经带动一批学生在数学学习当中以学习为主游戏为辅的方式进行学习，并得到了广大学生的一致好评！娱乐了乏味的数学学习，也调节了学习当中的枯燥情节，很多学生反映喜欢这种既能增加计算力又能提高理解力并能锻炼脑力的棋艺数学学习。

教学相长，何數棋谜中心的何校长在教学过程当中不断的与学生互动 利用教學機會更新教材，根据不同的学生級程度配制不同的教材.

何數棋谜所研發的材料最适合幼稚园到八年级之间的学生学习。如今，何校长已把此项发明创造发扬光大，在全世界建立了多家何數棋谜连锁培训中心.

何數棋谜是世界上數學棋藝綜合教學最大的連鎖教室

何數棋谜也成为当地各大华语报纸及华语电视台相继报道的"超便宜"加盟连锁机构。不仅如此，何校长的目标是让数学棋艺能够发展到全世界，让世界各族小朋友共同分享学习棋艺中的数学奥秘！

Student's name: ________________ Assignment date: ____________________

采访的最后，何校长真诚的希望，加拿大温哥华地区的华裔小朋友前来免费试用学习这项有趣的数学棋艺有趣味教学法！如果广大的家长朋友们还为孩子不爱学习而發愁，那么您不妨选择联系何數棋谜，联系电话：1-604-263-4321。

Student's name: ________________ Assignment date: ___________________

儿童数学幼教培训的重要

1. 何数棋谜是如何开始培训儿童幼教的？

由于何数棋培訓中心的创始人 Frank Ho 老师的儿子于 5 岁时对国际象棋发生兴趣引發了何老师研究国际象棋与数学关系的起源。而后何老师的太太 Amanda 师也加入了研發团队，結果是何数棋团队研發出世界独一无二有版权的数学，棋艺，谜题综合性教材, 劃時代新教材也震撼了儿童幼教界。中英文報紙, 電視台等媒介都報導過 (詳情見 www.mathandchess.com)。

2. 何数棋谜团队有什么样的背景？

Frank 何老师原来是由台湾得到成功大学统计学士，然后至美犹他及 BYU 大学获电脑学士及统计硕士，分别在犹他大学及 UBC 任统计顾问 15 年以上。何老师有 BC 数学老师资格证书。何老师的数学及棋艺論文也在 BC 數學專業教師論壇刊物發表過。Frank 何老师是兒童幼教專家。Amanda 老师原来是由中国大連理工學院得到理学士學位,到加拿大後 發篇教學論文 20 多篇, 並有 20 年以上教學研究及當工程師經驗。Amanda 老师是幼教及多科目教學專家。

3. 何数棋谜的幼儿数学培训有何特色？

何数棋针谜对资优，资平，资弱 3 种学生背景研发了超过 30 种数学的产品。并且成為全世界以游戏来教数学的领导地位。何数棋谜发明了教棋专用棋具，Frankho ChessDoku, Frankho Maze 等有版权产品。何数棋谜教材不但可以提高数学成绩，还可以提升脑力, 培養思維力。何数棋谜针对未來職場競爭要靠思考創新而研发出將数学,國際象棋,与謎題三大領域結合的教材，可以让 4 岁以上儿童在欢乐的環境中來学数学。

已成立 20 年的何数棋谜在 BC 有 4 家連鎖教室, 曾在私校 St. George's 開課, 成為温哥華小學生學數學棋藝及迷題人數最多,規模最大的培訓中心。何数棋谜在全世界有多家連鎖教室是一定有其獨到之處及讓眾人肯定的原因。

Student's name: ________________ Assignment date: ____________________

4. 儿童在幼年学数学是否非常重要？

根据科学报告及我们自己教学的经验可知，儿童在幼年把数学及逻辑的观念融会贯通是非常重要的。我们甚至可以追踪 10 年级的学生数学所以学不好原因之一是与小学的加减乘除計算能力未能熟練有关系。如果能在小朋友年纪小的时候提供機會來提升他们的脑力，加强论理的能力是十分重要的。

Student's name: ________________ Assignment date: ____________________

5. 何数棋谜培训中心与其他中心有何不同？

何数棋谜不但要提升学生的数学成绩，同时认为逻辑及脑力的提升对小朋友来说也同要重要，所以免费提供棋艺及逻辑训练的课程给所有报名数学的学生。何数棋谜独特研究的教材综合了数学，棋艺，谜题，奥数，文字应用题五大教学领域。丰富，多变化，及趣味的獨特高效率教材是其他培训该中心无法相比的。何数棋谜的教學法也巳經過科學統計試驗而証明有學前及學後顯着的差異。高年級生報名數學一科後，何数棋谜並免費同時提供多科目化學,微績分,物理教學。

Student's name: ________________ Assignment date: ____________________

怎樣避免投資一個燒錢的數學連鎖教室

在我超過 20 年的教學生涯中，我在加拿大溫哥華看見如此多的補習班開了又関。在過去十年中，有很多的教育中心建立，但是大多數都不見了，沒有地方可以找到它們。因此如果有人告訴你辦教育中心是一個非常容易經營的事業，不要相信它。

以為加入一個連鎖教室應該比開辦你自己的要容易也是一個錯誤。我對於這種想法的答案是消費者要當心。

有些"某某某"連鎖教育中心只是想收取你的連鎖費用而不關心你的成功與否，所以這是完全兩樣的事情。這種補習班的賺錢方式就是從連鎖費用當中盡可能的賺錢。例如，某某某連鎖教育中心要$25，000，它有 100 個加盟者，那麼這個教育中心能賺多少錢？讓我們來算算看：
$25,000 乘以 100 個加盟者=$2,500,000

對於主辦方來說一年就有超過二百萬的收入，不坏啊。對於加盟者來說會發生什麼？如果大部分加盟者在營業上掙扎，但是由於每年$25,000 的費用而不讓他運作，所以沒有勇氣把它関掉。這是一個很淒涼的事情。

你喜歡在生存期間漲了你 20%的租費又沒有正當理由的雇主嗎？沒有希望而且失望，誰需要這種生活？

不要把你的錢投資在一個不好的學習連鎖教室。買一個好的而且有規範的連鎖費用的教室。

很多教育連鎖中心給你一個很宏大的演講，並且反復說明他們的材料和教育方式有多好，但是如果你仔細觀察的話，你會發現他們當中的大多數描繪的材料有多完美是虛假的。

怎樣判斷它們的數學材料是與衆不同的？是題庫不一樣嗎？你可以在書店買到相似的題庫嗎？只是在數量上不同嗎？難道是題難一些但是對於孩子來說找不到一點樂趣的嗎？爲什麽材料可以提升孩子的思考技巧能力？有關於可以提升能力的研究報告嗎？

Student's name: ________________ Assignment date: ___________________

大多數的數學學習中心不能很清楚地回答以上的問題而都以用我們的材料是最好的，我們的學生喜歡它等藉口轉移問題而不接觸中心問題？

你應該知道在你支付了$25,000 之後買到了什麼樣的產品。這是對於買一個數學學習連鎖中心來說是最關鍵的問題。

Student's name: ________________ Assignment date: ____________________

如果學習中心的創辦者沒有任何特殊的教學思想，那麼我們怎能期待他/她能在十年后還能擁有一個學習中心？這只是一個玩笑。

學習數學應該以多樣化的方式而不是僅僅做計算的思想為前進方向。

數學學習中心是一個非常有競爭性的，並且只有最好的才能在時代的選擇中保留下來，所以不要加入一個沒有特殊成果的數學連鎖教室。買一所具有獨一無二的，有市場競爭性的並且有影響的數學學習中心吧。

Student's name: ________________ Assignment date: ___________________

如何選一個適合的數學培訓或家教中心

数學科培訓中心從經營的角度去分類可以分為“量產化”，“特定功能”，或“全腦單科”等三大類。各類皆有其優劣點，家長及學生需把各培訓中心創辦人的理念及經營的模式完全搞懂，這樣才能找到一個適合自己兒女學數學的培訓中心。

1. 量化式的培訓中心，這類補習中心在宣傳上也不會以產量化的言辭來吸引人的注意，但實際上只是大量的做計算題。上課只是去拿作業，作業做完即可走了。費用低以達到普及化是其優點，缺點是學生學數學最主要只是做做計算而已。
2. 特定功能式的培訓中心, 這類補習中心以只做數學競賽或專門只輔導學生學校數學功課為主。學生若不是數學實力本來就很強只做數學競賽就常有顧此失彼的感覺，因為學校功課不懂卻補不到。而只做學校功課為主即很可能只治標而非治本.
3. 全腦開發式的培訓數學中心，這類補習中心是學業及考試都兼顧的培訓中心。因材施教，因教施材，此培訓中心的特色是治標又治本，培訓不偏任何一方。數學能力是全面性的提升，包括學校成績，數學競賽及腦力等的同時提升。何數棋谜更結合了數學，國際象棋及謎題，所以有獨特研發的材料，使學生對數學感到更有興趣。缺點是這種培訓中心的創立門檻高，所以不容易開設. 但何數棋谜有研發團隊，每年都有新創教材及作業，所以在所有培訓中獨樹一格，走出了自己的路。全世界能夠真正結合數學，國際象棋及謎題而只付需一科費用，多科目學習的培訓中心只有溫哥華龍頭培訓中心 – 何數棋谜一家.

Student's name: ________________ Assignment date: ___________________

介紹何数棋迷宮數學™ – 寓教於樂

何老師為世界首創數學棋谜有版权教材發明人及

何數學棋谜創辦人

1. 什麼是何数棋迷宮數學™?

今天兒童面對的世界是學習如何綜合處理數字,圖形,資料搜尋,音影上下載,資訊比較,分類等資訊.這些活動實際已成為兒童生活的一部份.所以如果說學數學就是學數字就錯了,學數學的另一個目的就是學習如何解決問題及培養創造力.但是傳統式數學的計算練習題卻完全沒跟上科研已經改變了兒童面對的世界. 兒童想要的計算題已經不是單純的從上到下,從左到右的純計算.兒童需要的是他們情願的而又快樂地做不枯燥的計算題.

如何將傳統式數學計算題變得有趣而且又好玩,並且還可以增強兒童的計算及解決問題能力及培養創造力,同時還可以增進兒童記憶的能力達到全腦開發的目的?

何老師首創已申情版权的幾何棋藝符號並利用此符號發明了世界第一的數學棋藝專利教材.何數學棋藝讓兒童能利用幾何棋藝符號進行邏輯推理及數字的運算.棋藝與算術的綜合題含蓋了整數,幾何,集合,抽象數,對比異同,函數,座標,多空間圖形資料,及規則性數字分析.並且把棋藝的趣味性和數學的知識性結合在一起.

Student's name: ________________ Assignment date: ____________________

何老師研究的專長是如何將棋藝融入數學以達到寓教於樂的教學理念. 何老師研究的棋藝是西洋棋或稱國際象棋. 國際象棋本身就包含了許多數學的觀念例如記錄每步棋的代數符號記錄法, 將軍的集合數學觀念等.

Student’s name: ________________ Assignment date: ___________________

這些數學的觀念並沒有包含在國一或國二學校的數學教材. 因明顯的小學生下棋時正使用他們沒有學過的數學觀念. 何老師發明的數學棋藝專利教材就是很明確的將棋中包融的數學歡念抬上桌面讓小學生明白棋中數學的觀念.

世界上許多人都被如何將棋與數學結合在一起而困惑了. 許多人也曾研究過但都沒有突破. 何老師能成功的將國際象棋融入數學就是因為他發明了幾何棋藝符號 (專利權申請中). 這是一把鑰匙解開了數學與棋藝連接的奧妙.

數學與棋藝的關係可以從三個大方向來分析:

1. 每一個棋子都有一國際規定的點數, 例如王后是 9 點, 小兵是 1 點. 這些棋子可以說本身就具備了代數的關念.
2. 每一個棋子的走法都代表了幾何線, 線段, 及交叉點的概念. 例如城堡可以上下左右移動.
3. 下棋的思維方式從觀察全盤布局, 分析未來走法, 到下結論都與數學解題的一般原則完全一樣.

何老師成功的以計算題為平台,然後輔以訓練思維方武的思考題(幾何棋藝符號巳申請專利)為仲展台以加強學生不但可以增強計算能力並且還可以增強解題能力及培養全腦開發創造力.

Student's name: ________________ Assignment date: ____________________

2. 如何教學?

何數棋谜的教學法並不是將數學與國際象棋二科目分另別放在一個教室來分別教學. 我發明了幾何棋藝符號, 所以成功的將數學與幾何連接在一起.
我也利用幾何棋藝符號發明了專門教小朋友下棋用何數棋谜教具棋. 非常顯著的效果是兒童不再為每個棋子的走法而困腦.四歲的兒童經過介紹幾何棋藝符號後可以立刻使用何數棋谜教具棋下棋.

何數棋谜發明了整合棋藝與數學的專利圖形語言幾何棋藝符號,讓兒童能利用符號圖形進行邏輯推理及數字的運算.棋藝與算術的綜合題含蓋了整數,幾何,集合,抽象數,對比異同,函數,座標,多空間圖形資料,及規則性數字分析.並且把棋藝的趣味性和數學的知識性結合在一起.

我們的教學理念是希望兒童能浸浴在一個歡樂的環境下來快快樂樂的學數學,讓數學來適合幼童的興趣而非強迫幼童做題.

綜合言之, 何數棋谜 有系統地研發了融合數學與國際象棋的數學教課程而且以達到 寓教於樂, 按個人進度個別施教 的教學理念.

3. 學數學棋藝的好處在那裏?

上百篇科學論文已發表國際象棋可以提高兒童英文及問題解答能力.並且訓練他們的專心及耐力.所以我們已經知道下國際象棋對兒童有好處.但是因為國際象棋與計算能力並無直接開係,所以如何讓兒童能在一個歡樂的環境下也能利用下棋來提高數學的計算呢?

對許多學前兒童到國小四年級生來說, 數學是個枯燥沒有樂趣的科目. 因為每個學生說到數學只是在做計算題吧了. 我成功的將國際象棋融入數學, 使數學與棋藝的結合達到了相輔相成的效果, 何數棋谜讓兒童不但達到了學棋的好處學而且也同時在寓教於樂的理念下增強數學的能力.

Student's name: ________________ Assignment date: ___________________

寓教於樂的教學法是最有效的學習數學法.科研報告已經證實何數棋谜的教學法不但可以提高兒童數學解題及思維能力,還可以開發兒童的腦力,及分析問題的能力並且增加兒童學習的耐力,訓練機警靈巧及加強手`腦`眼的靈活運用.

Student's name: ________________ Assignment date: ____________________

4. 如何幫助兒童腦力開發?

很簡單的一個道理就是讓學生自願地去用腦, 何數棋谜首創獨一無二的融合數學與棋藝的專利教材,利用棋藝訓練右腦的座標,空間分析及圖形處理,並利用發明了整合棋藝與數學的專利圖形語言幾何棋藝符號,讓兒童能利用符號圖形訓練左腦進行邏輯推理及數字的運算.棋藝與算術的綜合題含蓋了整數,幾何,集合,抽象數,對比異同,函數,多空間圖形資料.所以枯燥無味的計算題變成了謎題,學生需要通過更多的思考, 能讓腦去思考愈多則腦力也愈開發.處里訊息,分析資料才能發掘出題目.學生在做這些謎題式數學時比較會專心及有耐心.

5. 小朋友是否對象棋要有興趣?

答案是否定的. 因為家長或兒童如有先入為主的觀念認為下棋沒有必要與數學合在一起,則對我們的理念來說反而有阻礙. 興趣可以培養, 只要家長與兒童願意接受新觀念接受挑戰則成功的機率很大.

6, 初學小朋友會遇到什麼困難? 如何克服?

有些小朋友會認為我們的教材比較難而向家長抱怨, 心軟的家長因而打退堂鼓宗. 所以應因之道是小朋友需要家長及老師更多的鼓勵及小服友的加倍努力.

7. 當初是如何開創學數學棋藝?

我的兒子(Andy)於 5 歲多時對國際象棋發生興趣,但我卻不懂國際象棋,於是去圖書館借書,并且以電腦為教學工具,展開了一連串的父教子的自學活動 兒子於 12 歲時得了加拿大 12 歲組,少年(14 歲)組,青年(18 歲)組的三冠王,

破了加拿大棋界的紀錄,並且於巴西的分齡世界國際象棋大賽時得第 5 名.

兒子為 FIDE 國際象棋大師.

許多人都知道數學與國際象棋有著密切的關係,但市面上并沒有專門書介詔小朋友國際象棋與數學的關係何在.就是因為這個自已教兒子下棋的機緣,我對國際象棋與數學的關係發

Student's name: ________________ Assignment date: ___________________

生興趣.於是在 1995 年我發明了世界上獨一無二,融合數學與西洋棋的教材.並且創辦了讓學生有個快快樂樂學習環境的獨特的而且為世界上唯一的融合數學與國際象棋的何數學棋藝™培訓學習中心. 何數棋谜讓學生在歡樂的氣氛下增加學習數學的興趣.

Student's name: ________________ Assignment date: ____________________

我融合數學與國際象棋的教學理論已在 BC 省數學教師刊物上發表.

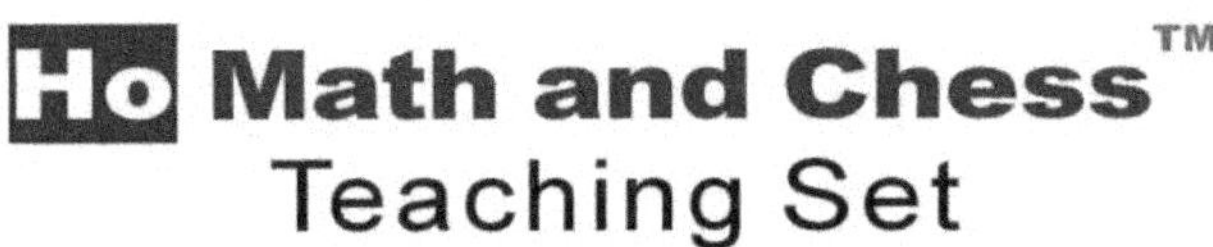

- Specially designed for children.
- What you see is what you move.
- Is easier for children to learn chess.

www.mathandchess.com Patent pending.

⚠ Warning: CHOKING HAZARD-Small parts not for children under 4 years.

Made in China

box design

Student's name: ________________ Assignment date: ____________________

兒童神奇數學、棋藝、腦力、潛能開發

何數棋谜™ 的創辦人何老師(Frank Ho)為世界棋藝與數學教學的權威.他所發明的棋藝與數學多功能綜合教材及教學專用棋子都已申請了專利.

何數棋谜™發明了整合棋藝與數學的專利圖形語言,讓兒童能利用符號圖形進行邏輯推理及數字的運算.棋藝與算術的綜合題含蓋了整數,幾何,集合,抽象數,對比異同,函數,座標,多空間圖形資料,及規則性數字分析.並且把棋藝的趣味性和數學的知識性結合在一起.

我們的教學理念是希望兒童能浸浴在一個歡樂的環境下來快快樂樂的學數學,讓數學來適合幼童的興趣而非強迫幼童做題. 何數棋谜™寓教於樂的教學法是最有效的學習數學法.科研報告已經證實何數棋谜™的教學法不但可以提高兒童數學解題及思維能力,還可以開發兒童的腦力,及分析問題的能力並且增加兒童學習的耐力, 學生的探索創造精神及求知欲.判斷力,及
自信心等. 啓發思維訓練機警靈巧及加強手`腦`眼的靈活運用.

Student's name: ________________ Assignment date: ____________________

Student's name: ________________ Assignment date: ___________________

腦力廝殺明報記者報導 (非廣告)

Chinese Ming Po Report (non-advertisement)

相傳在大約兩千多年前，在印度曾經發生過一場激烈的戰爭。戰爭過後，尸骨成山，血流成河，真是慘不忍睹。有人眼見這種景象，立即做了一塊有六十四格格子的正方形棋盤，塑造了一些形態各異、戴盔披甲的戰士做為棋子。他把戰場上的戰鬥再現在棋盤上，終於把恃強好勝的國王、將軍及婆羅門貴族的興趣吸引過來。從此在棋盤上較量智力，取代了戰場上的血腥廝殺。

恩怨愛恨世事如棋—每局都充滿傳奇

姑且勿論傳說的正確性，可以肯定的是國際象棋在這麼多年來，吸引了很多人，亦成為訓練兒童思考的好教材。國際象棋又名為西洋棋。在棋藝的世界中對戰雙方都擁有相同的兵種，在平等的條件下，如何將對手打敗，則視乎棋手的想像力、判斷力和耐久力，每一步都是勝敗的關鍵所在，絲毫大意不得。
有很多研究證實國際象棋對兒童的思考方式有幫助，Dr. Robert Ferguson 曾經用 4 年的時間，以第 7 至 9 班的學生做實驗對象，發現沒有玩國際象棋的學生在批判思考 (critical thinking) 方面的能力，平均每年增強百分之 4.6，而學習國際象棋的另一群學生則為百分之 17.3。何數棋谜 (Ho Math Chess Learning Centre)的創辦人何鳳台表示："學習國際象棋並不能保證小朋友在學科成績上有進步，他們學到的是思考方式。"

何數棋谜中心

何鳳台的兒子何其佳 5 歲的時候，收到一份朋友送的國際象棋。意外的是，小其佳竟深深地被它吸引。父子倆人便一起鑽研學習，後來其佳在 12 歲的時候獲得加拿大 12 歲組、少年組、青年組的冠軍，破了當時加拿大棋界的記錄，并於巴西的分齡組世界國際象棋大賽比賽中獲得第 5 名。

何鳳台對象棋和數學之間的關係產生了濃厚的興趣，於 95 年在溫哥華開辦了一間數學和棋藝的學習中心，讓學生在課餘的時候，用有趣的方法學習數學。在研究的過程中，何鳳台發現市面上相當缺乏適合小朋友的參考資料，他便開始自己編寫了一本專為第一班或以上的學生而設的教材--《小兒數學數學棋藝》(Mathematical Chess Puzzles for Juniors)，內容包括有超過 100 個棋謎。
要解答書中的謎題，不但要對國際象棋有基本認識，還需要運用到課堂上學到的數學知識。

Student's name: ________________ Assignment date: ____________________

張眼遠望世事如棋 — 每局應觀察入微

國際象棋是科學、文化、藝術、競技融為一體的智力遊戲。它有助於開發智力，培養邏輯思維和想像能力，加強分析能力和記憶力，提高思維的敏捷性和嚴密性。國際象棋的棋法多變、趣味橫溢，很容易引起小朋友興趣，自發性的要求學習。以 12 歲的 Eric Lerer 為例，他在 5、6 歲的時候已經很喜歡玩象棋，後來他希望能夠改進自己的技術，所以才正式上課學習，年紀雖輕，但已參加過很多公開比賽，如卑詩省象棋錦標賽。

另一位學生熊本寬的母親亦表示："有一次我看到他跟其他小朋友在玩國際象棋，他只是第一次接觸，便玩得很好而且很有興趣，所以便讓他學習，看看如何，並不是要他在這方面有什麽成就，只是希望能協助他發覺他自己的嗜好。"

國際象棋對於開發少年兒童的智力，特別有好的效果，而且對閱讀等方面都有所幫助。因此，目前世界上已有不少國家把國際象棋列入小學課程。雖然加拿大仍未把國際象棋歸納為正式課程的一部份，家長不妨讓小朋友試一試這種有趣而有益的遊戲。

Why buy a Ho Math Chess Franchise?

Intellectual Properties of Ho Math Chess

何数棋谜 - www.homathchess.com

Student's name: ________________ Assignment date: ________________

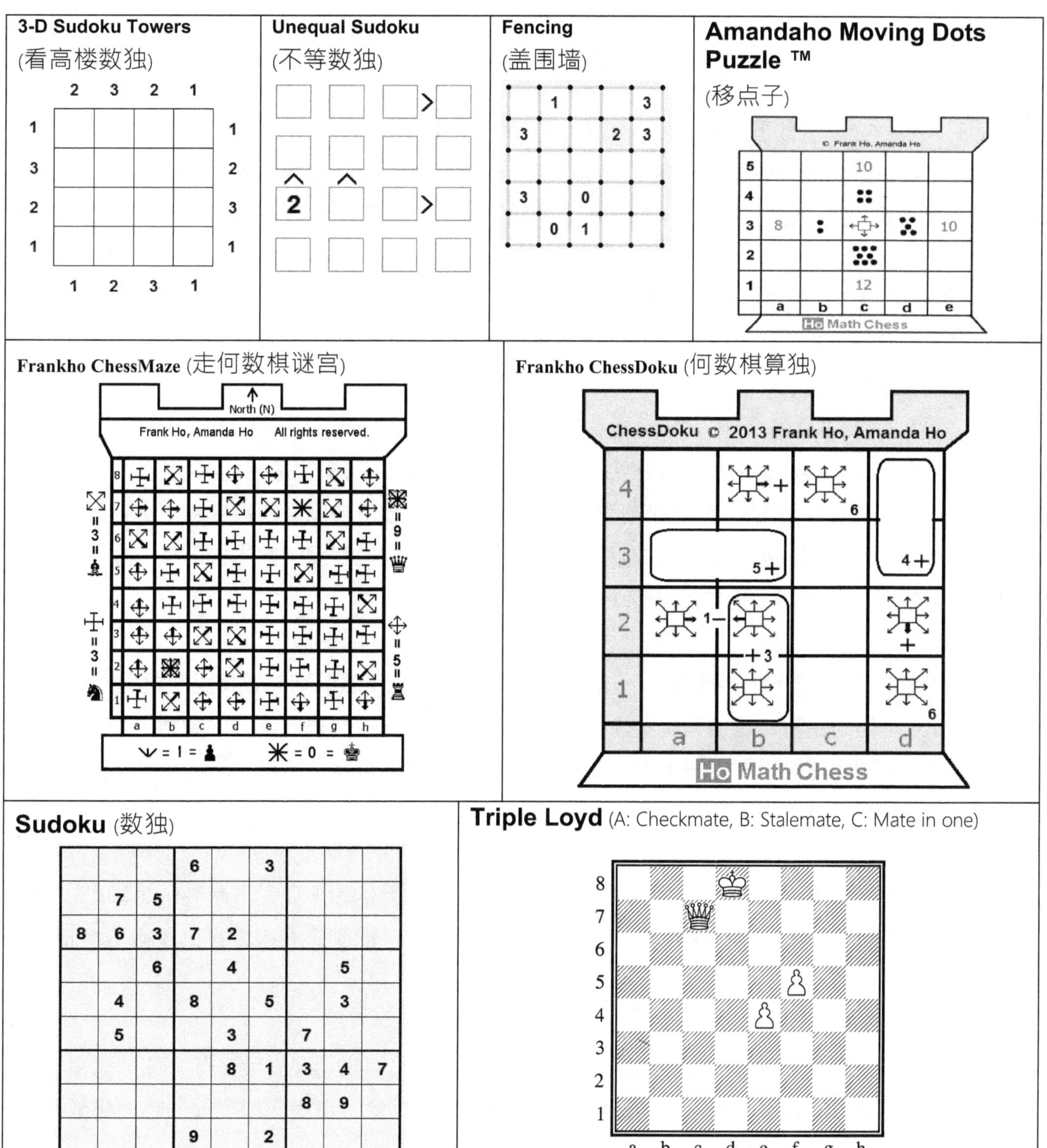

Student's name: ________________ Assignment date: ____________________

我儿子怎样成为象棋大師

我的兒子 Andy 12 歲就成了加拿大青少年象棋冠軍、國際象棋協會頒佈的大師和加拿大象棋大師，打破了加拿大棋界的紀錄。十几年后，仍不断有人向我打听，到底怎样培养儿子为国际象棋大師。我们父子走过的是一条与众不同的路。我把它写下来供大家参考。

Andy 五歲對國際象棋產生了興趣，当時我對國際象棋還一無所知。我想，如果能和他一起學習國際象棋，將是非常有趣的事。于是我去溫哥華圖書館借了些有關國際象棋的書，開始用這些參考書教他怎樣下象棋。

当 Andy 學会了象棋的走法，我們就开始一起玩象棋。Andy 最初學的是殘局。初期，在教他之前，我不得不自己先學會各種各樣的殘局戰略，但很快我就發現自己學的進度趕不上他的吸收速度。于是不得不采取另一种方式，就是我唸書他下棋，让他跟據我讀的書本去下棋，通常我读完自己并不甚解，他却理解了書本的含義。

Andy 在幼稚園學棋僅僅幾個月就贏了一年級的前冠軍。这时我意识到需要找比我更懂棋的棋手來教他。90 年代的溫哥華還沒有少年象棋俱樂部，我不得不帶他去成年人的象棋俱樂部。可是，许多成年人不習慣和這麽小的孩子玩棋，排斥他。這促使我以後自己創辦了何數棋谜（Ho Math Chess）。

当年象棋俱樂部有一位退休醫生有極大興趣和兒子下棋，那醫生和下兒子下棋時，每移動一步棋子都用很长時間（那時沒有用計時器）。大概因為這樣，兒子剛剛過 6 歲便被訓練得下棋時顯得十分有耐心。

很快我又發現，由於与其他孩子的棋力已拉开檔次，Andy 已有棋不逢對手之感，急需繼續提高棋力。卑詩大學(UBC)每周二晚上的棋藝競賽是一個很好的机会，自從參加了周二晚上和成年人的對戰，Andy 獲得很多經驗。

Student's name: ________________ Assignment date: ___________________

記得當年 Andy 代表加拿大參加 10 歲國際大賽，他的棋力是 1900 左右，但國際上其他對手有些已達到 2000 的水平。這次國際比賽的經历讓 Andy 和我受到極大的震撼，我們对後来的訓練做了重新調整，購置(市面上剛出現)手提电脑、建立開局分析報告、整個開局的强勢出局、戰後的分析及檢討这些都是从國際比賽得到的寶貴經驗。

國際賽後回加拿大，最困擾我的是找不到一個教練可以教導 Andy，把他的棋力提升到新的高度甚至世界水平。我發現，怎樣把一個小孩子訓練登上國際舞臺，多數教練都沒有經驗。對 Andy 來說，從書本中學習可能是最好的選擇。

在接触了幾個象棋私人家教後，我找不到可以使 Andy 繼續進步的教育方式。一天我突然想到，象棋大師的棋譜都是公開的，為什麽我不分析他們的棋譜，让 Andy 學習他們開局的方式呢？事实證明，這種無師自通的方式對 Andy 來說是可行的。

一連串的父教子的自學，让兒子 12 歲时得到了加拿大 12 歲組、
少年(14 歲)組、青年(18 歲)組的三冠王，破了加拿大棋界的紀錄，並在巴西的分齡世界國際象棋大賽獲得第 5 名，兒子成了 FIDE 國際象棋大師。
而日夜專研棋的代價也讓我深深體會古時拜師三跪之禮的意義。90 年代我是怎樣在沒有教練的情況下，把 Andy 訓練達到世界水平呢？主要從以下幾方面着手：

1、与幾個退休的象棋大師聯係。

2、學習中國人如何訓練少年棋手，分析一些象棋大師是怎樣被訓練的。在網上尋找其他國家怎樣訓練少年棋手的資料。

3、添置象棋錄影帶和如何開局的象棋書籍，訂購世界上所有主要的象棋雜誌。

4、添置電腦象棋軟件和個人電腦，全天候的練習棋藝。

Student's name: ________________ Assignment date: ____________________

做完以上資料分析後，我的結論是，比賽時，如果棋手在開局獲得先手，在心理上就已佔了上風。如果在開局理論上成為專家，一個平常水平的棋手就很难打敗一個專家棋手。我把钻研開局理論作为 Andy 的重點訓練項目。這種無教練訓練方式最大的好處就是可以在任何時間、任何地點進行，所需要的只是一台電腦和象棋書籍。

訓練 Andy 時我遇到了另一難題。我發現當遇到不熟悉的開局時，棋手們花費太多時間去尋找答案。90 年代初，我利用當時兩本最流行的象棋開局書籍研究學習了所有主要的開局，为 Andy 找出遇到不同的開局時可以運用對應的各種棋路。我們把他所有的對應開局走法画在圖紙上，最後這些圖大到可以覆蓋住整張辦公桌的表面。

Student's name: ________________ Assignment date: ___________________

Andy 的腦子裏有個記憶庫，他的對手走什麼棋路时他都知道如何回應。他對於有關 "What if" 的開局方式準備得很出色。如果對手下的棋路是他喜歡的開局，那麽大多數棋手都沒有能力去反擊。有一段時間 Andy 在溫哥華甚至被稱為是 "終結者" (Terminator)。

Andy 很快就成了最小的加拿大青少年象棋冠軍、國際象棋協會頒佈的大師和加拿大象棋大師。我的訓練方式也许可以訓練一個孩子成為象棋大師，但我也相對付出了極大的代價：往往清晨 2 點還在研究第二天的戰法，日夜不休的研究、無限的時間投入及極大的擔憂、心力及財力的巨大花費。

我想过，如果時間能倒轉的話，我也许不會再花費如此多的時間及心思去訓練 Andy。作為替代，我會挪用一些下棋的時間去指導他的數學知識。因為數學會幫助 Andy 一路通往大學甚至研究所。這也是我創立何數學棋藝學習中心，把數學和棋藝聯合起來教授的原因，這樣，孩子們就能同時學習數學和國際象棋了。

注：作者为加拿大數學資格證書教師，何數棋谜創辦人。家長見證、數學研究報告、課程作業樣本等，請查詢：www.mathandchess.com。

Student's name: ________________ Assignment date: ____________________

下國際象棋真的可以培養數學能力

如果說會下國際象棋的結果是學生數學一定好是個真理，那麼爲什麼全世界各國不把數學改成下棋課呢！所以說數學能力提高並不一定是下國際象棋的自然結果. 但是我們也可以由許多國家是提倡下國際象棋課得知國際象棋與數學有確實有相當大的關係，何數棋謎的網址 www.mathandchess.com 有篇何老師發表的國際象棋與數學關繫的文章，家長學生可以上網查詢。

國際象棋與數學的關係到底是什麽，下國際象棋真的可以提供數學能力嗎？根據我們教國際象棋與數學15 年的經驗，下國際象棋是真的可以提供數學的能力，但是其做法必須是有下列條件：

(1) 國際象棋可以提升數學應用問題的思維能力，但學生必須認真的專心學習國際象棋，而非抱著只"玩" 遊戲的態度。這就是爲什麽何數棋謎開發了數學,棋,及謎題的教材，學生必須要有作業而且做數學,棋,及謎題的教材，學生不只是要有實際 "作戰" 的經驗，" 紙上談兵" 作業的功夫也不可缺。

(2) 國際象棋必須與數學緊密結合，否則二者的關聯性不能被學生很容易的體會出來。何數棋開發了棋，數學，謎題三結合的教材,

這些教材可以提升學生的腦力與計算的能力。國際象棋不能與數學結合，其結果只是讓學生覺得國際象棋只是種遊戲而巳，而許多學生還不喜歡下棋。在此倩況下, 國際象棋不能與數學結合, 效果就不好.

何數棋學生尚可以做 "數棋" 材料而不下棋，同樣的可以達到練腦力的效果。 "何氏智能數棋謎題" 及 "何氏益智迷宮" 等教材的發明就是利用國際象棋來訓練學生的智能而提升腦力，但學生並不要下棋。

下國際象棋而能提高數學解題能力最主要的原因是因為有些棋子形成的規律是數學上根本看不見的。如 pin, fork, skewer, discovered check 等等這些下棋的技巧可以說是特殊的因果規律，這些規律只做數學是不可能學到的。而下棋用的策略更是學數學只講 "倒轉法" ， "畫圖法" ， "估算法" ， "工程問題" ， "雞兔問題" 等解題策略不可能學到的。下棋策略的運用可以讓學生更增加對人生歷練的瞭解，而且更接近社會的真實情況，因為對手在每一步棋都要想破解你的攻擊，對學數學的解題策略來說，這簡直是不可思議的，解數學題是不可能每一步都有另一個對手來阻擾你的解法的。

Student's name: ________________ Assignment date: ____________________

設想周到，反應靈敏，頭腦靈巧，眼觀八方，步步為營，心靜如水，控制情緒都是可以從下棋中學到。這些下棋的好處都會化為烏有，如果學生不是在有方向的引導下認真的去學習，因為懶散的去學，其結果只是“好玩”罷了，其結果是不能提升數學應用問題的思維能力。

Student's name: ________________ Assignment date: ____________________

能夠在下棋是認真的，嚴肅的去思考，走一步之前先想一想，結束後再檢討，在這種認真學習的態度下，下棋的好處就會展現出來。如果再能將國際象棋，數學，及謎題結合在一起，學生也可以享受到因學棋及做謎題而得到好處，何樂而不為？數學成績提高了，腦力也提升了，做謎題及下棋的樂趣也培養出來了，這些好處可以影響到下一代，因為這些小朋友將有一天也為人父母，同樣的，他們會把數學，國際象棋，及謎題的樂趣及好處傳授給他們的下一代，一代傳一代，腦力也一代比一代提升的更高了。

Student's name: ________________ Assignment date: ____________________

以国际象棋與謎題作為數學遊戲教學實際經驗的報告

本文不探討遊戲教學是否能激發學生對學習數學的興趣，因為我們知道它的答案是肯定的。

本文是報告以國際象棋與謎題來教育學生 15 年來的實際數學教學經驗。

一般以數學遊戲教學作為單獨課題實驗研究的缺點如下：

- 一般以不同類型的數學遊戲來做遊戲教學都以數個月等短期做實驗研究，所以短期研究的結論是否有長期的效果（如數年）仍待研究（林嘉玲 民 89）
- 數學研究單項教學常以某一特定年級學生為對象，以其結論能否適用于各種不同年級仍有待研究。
- （林嘉玲 民 89， 葉盛昌 民 92 等）
- 數學研究都是單獨遊戲作業而未正式與數學作業融合（王克蒂 民 88），所以不論教學的方式是融入於正式課程或以數學團隊方式教法，其方法都與何數棋謎題將數棋謎與數學作業正式結合的方式不同。

如何改進以上數學遊戲的缺點呢？

何數棋謎趣味教學中心以國際象棋及謎題為遊戲工具，並將它們正式與數學作業結合供幼兒園及

國小生(一年級至七年級)長期使用以達到學生在歡歡樂樂的環境下學習數學的效果。

為何使用單一遊戲國際象棋呢？

國際象棋有挑戰性，玩的精的學生甚至可以成為象棋大師，具有極高榮譽的挑戰性，而棋的走法千變萬化，所以不是公式化。如果不小心下不好棋，則對手很可能會處罰。更

Student's name: ________________ Assignment date: ____________________

妙的是，對手會想盡辦法不讓另一方達到目的，所以此教學遊戲極具競賽性。而對手下錯棋即可能給另一方製造了勝的“機運”。

國際象棋不是學生在小時才可以玩或自己個人玩的遊戲，當學生 80 歲時如有興趣仍可以繼續玩并訓練腦力。所以國際象棋不會成為一種一成不變，制式化的遊戲，玩的愈好，興趣也愈高。此點有別於饒見維（民 85）必需多種數學遊戲的建議。

何數棋謎趣味教學中心將國際象棋與數學的練習題融合，所以學生可以將數學當遊戲式的做練習，不見數學題目，但見謎題，初看還真讓人墜入了如雲霧之間，而使學生迷惑，卻激發了學生的好奇心。

何數棋謎將國際象棋，謎題，與數學傳統練習題正式的結合開創了另一種數學遊戲教學法，也將數學遊戲帶入了另一個新的定義，即數學遊戲教學法不再是只是短期的讓某一特定年級的學生以遊戲的道具做些數學題來增進學生學習數學的興趣。

何數棋謎培訓中心對數學遊戲定義為將國際象棋，謎題正式與國小數學練習題融合在一起，使學生自然地認為國際象棋迷題就是數學的一部份，所以國小學生上數學課就是在上數學，國際象棋及謎題的課。而何數棋謎培訓中心的教材除傳統式計算題外，也就包含了“數棋謎”三合一的教材。

林嘉玲（民 89）：數學遊戲融入建構教學之協同行動研究。國立花蓮師範學院科學教育研究所碩士論文，
未出版。

葉昌盛（民 92）：遊戲式數學教學模式對學生數學學習的影響。國立臺中師範學院數學教育學系在職進
修教學碩士學位班碩士論文，未出版。

Student's name: ________________ Assignment date: ____________________

王克蒂（民 88）：數學遊戲教學之效益研究—以國小四年級學生為例。國立臺灣師範
大學科學教育研究
所碩士論文，未出版。

饒見維（民 85）：國小數學遊戲教學法。台北市：五南。

Student's name: ________________ Assignment date: ____________________

World Junior Chess Championship

By Frank Ho

In mid-July, as a parent, I accompanied my son Andrew Ho to participate in the World Junior Chess Championship held in Bratislava, which is now the capital city of the Slovakia Republic. There were five young Canadian players competing in the world event. The final standings were as follows:

Stephanie Chu, Category G-10, 6.5/11, #12 of 44.
Andrew Ho, Category B-10, 6.0/11, #20 of 50.
Lefong Hua, Category B-12, 5.5/11, #32 of 68.
Anthony Castillo, Category B-14, 5.0/11, #46 of 72.
Umadesan Casinathan, Category B-16, 4.0/11, #56 of 75.

The trip offered a very valuable and rare experience for Andrew, which hopefully will help him if he has a chance to go again. As a parent, it provided me with a precious opportunity to talk to other parents from different countries, discover some insights, and exchange training ideas.

Since the results were far from the high expectations anticipated before the competition, it inevitably led me to ponder a series of questions after the competition was over. Although the Canadian team was a small contingent compared with some teams who had a representative in every category, it was by far the largest and strongest team ever presented in the world youth event. All participating Canadian players are champions in their own age groups in Canada. The results show that the top participating players' level in other countries are generally higher than the same age group of Canadian players. Why is it? I do not know the exact answer. However, by talking to players' parents and coaches and by my personal observation, I think there are a number of reasons and I would hope the information provided here will be helpful for young chess players and their parents to better prepare themselves to face the world challenge in the future.

When compared with other top players, especially European players, Canadian players do not have as many chances to participate in chess events with international participants, probably due mainly to geographical location. The lack of chances to play with top young players may have placed Canadian players in a disadvantaged situation. Fortunately, young players could overcome this by consistently playing in the open or higher sections to improve skills.

Some of the top players are trained in a very structured and formal way (such as going to "chess school") and chess players selected are very talented and dedicated. In contrast, all five Canadian players play chess as one of their hobbies. To compete in chess on top of other activities, it means Canadian players would have to use time more wisely in order to compete with other players who practice two hours a day to prepare for the world event. This could mean making use of the 20 minutes while mom or dad is shopping or whenever there is free time.

The opening lines used and all variations must be thoroughly understood, and this does not mean the players just play the book lines - creativity and ability to change lines are the training goal for winning. Players should also keep their eyes on any new opening lines or theories by subscribing to magazines, or purchasing new books.

To have good results in the world event, it means the winning chances for playing either the Black or the White side should be at least 80% or better. Using this year's players' strengths as a standard by which to predict the future, I would say that a player who consistently has good results when playing in the 2000+ section could expect to do well in the World under 10 boys. Players who compete in the World under 12 boys may expect to do well if their rating is around 2300. The rating indicates how tough and strong a player should be when competing in the world event. Apart from the training for players, the planning for the trip should start as soon as the player has the potential to participate.

Lots of our friends in Vancouver have helped Andrew's trip in one way or another, our hats tip off to all of you. B.C. Junior Chess Co-ordinator Mr. Harold Daykin's support from the beginning of the trip to the end is greatly appreciated. Last but not the least, Andrew and I would like to take this opportunity to thank the Chess Federation of Canada, the B.C. Chess Federation, the Lucy L. Woodsworth Fund for Children, and the Taipei Economic & Cultural Office for their financial support for Andrew's airfare to Bratislava.

1994 Northern B.C. Winter Games

The 1994 Northern B.C. Winter Games will be held in early February 1994 in Fort St. John. After last year's successful chess events in Quesnel with 40 players, we are hoping to attract a large number of adult and junior chess players over the age of 12. Gold, silver, and bronze medallions will be awarded in four categories - 12 to 15 year old juniors, 16 to 18 year old juniors, rated adults, and unrated adults. The two junior sections will be CFC-rated and the winners may qualify for financial assistance from the Northern B.C. Chess Association to travel to Vancouver to play in the provincial Chess Challenge Championships by grade to be held in March 1994. In order to qualify for your zone in Northern B.C., you must contact your local sports co-ordinator for the Games and play in a qualifying tournament if it is required. Entry forms for the Games sports must be completed by the end of November; so hurry, and contact your co-ordinator. Bus transport to the Games is subsidised and all junior participants are billeted by local families. There are also social events for juniors in the evening and the young people have a great time!

If you have any further questions, please call Jim Kanester at 782-5637, or Larry Stutzman at 785-7830.

Notice

Counterplay has now changed publication dates from even-numbered months, to odd. The magazine will be sent out the week before the month of publication. In order to meet this goal, new submission dates have been set. All submissions other than camera-ready advertising should be received no later than the 1st of the month before the issue it is being submitted before. Camera-ready advertising will be accepted up until the 8th of the month. For further details on the requirements, see the sidebar on page 2. Thank you for your patience.

Student's name: ________________ Assignment date: ____________________

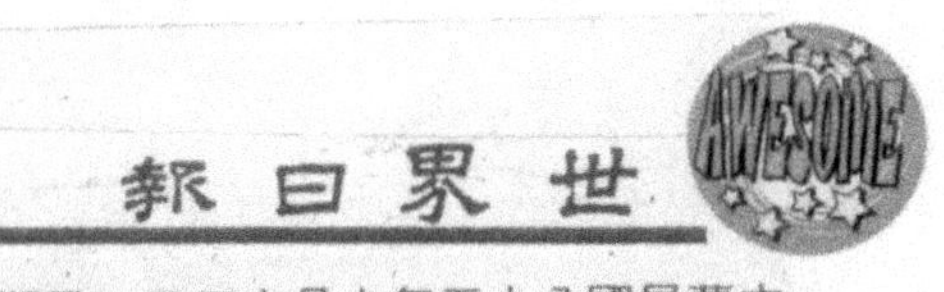

世界日報

中華民國八十五年十月十三日 星期日

華裔小棋士 月底將代表加拿大赴西班牙參賽

何其佳 問鼎世界青少年棋王

【溫哥華訊】這個周末在本那比卑詩理工學院內舉行的「大師級感恩節棋賽」（Thanksgiving Day Masters Tournament）中，好手如雲，如一九九四年美國公開賽冠軍奧洛夫（G. Orlov）及最近才獲得加拿大錦標賽亞軍的泰皮茲基（Y. Teplitsky），但當中最受矚目的可能要算是十三歲的華裔小棋士何其佳。

目前就讀於 University Hill 中學的何其佳六歲開始學棋，迄今只有七年，但已是加國年輕一代最被看好的棋手。他在去年曾代表加國赴巴西參加世界青少年賽，榮獲第五名，本月十九日他將再次代表加拿大前往西班牙，參加這項世界棋賽十四歲級（Under-14 Division）的比賽，因此這兩天的比賽可說是他的暖身賽。

卑詩理工學院感恩節大師賽主辦人布魯斯特表示，雖然何其佳年齡較小，卻不能因此就認定他會在這項好手雲集的大賽中遭到痛宰，「他不一定能贏得冠軍，但是誰敢說他不可能與國際大師級的選手打平呢？甚至贏棋都有可能。」何其佳自己也表示，對他而言，本周末的比賽是在世界大賽前很好的準備賽。

何其佳的父母都是台灣移民，他的父親何鳳台在一九七八年來加，原本在卑詩大學擔任統計顧問，因健康因素目前已經半退休。何鳳台說，七年前他帶著其佳一起去學棋，沒想到自己進步有限，其佳卻進步神速，逐漸展露他在這方面的天份。

何鳳台說，其佳目前不定期接受泰皮茲基的指導，但主要的學習管道還是在家中與電腦對弈，此外他們也為其佳訂閱了三份棋藝雜誌。

何其佳去年即已完成所有七至十年級的課程，目前在 University Hill 高中修習十一年級的課程，他在課餘也彈鋼琴（目前已是第十級），還自行學習設計國際網路上的網路本頁（Homepage），他希望不久後即能完成這項設計工作。此外，他還練柔道，不過由於課業、下棋、電腦等已佔去太多時間，柔道已經停下一陣子未練了。

www.ingramcontent.com/pod-product-compliance
Lightning Source LLC
LaVergne TN
LVHW081402110826
845149LV00010B/1638

* 9 7 8 1 9 8 8 3 0 0 4 9 8 *